T0220464

Artificially Induced Grain
Alignment in Thin Films

MATERIALS RESEARCH SOCIETY
SYMPOSIUM PROCEEDINGS VOLUME 1150

Artificially Induced Grain Alignment in Thin Films

Symposium held December 2–3, 2008, Boston, Massachusetts, U.S.A.

EDITORS:

Vladimir Matias

Los Alamos National Laboratory
Los Alamos, New Mexico, U.S.A.

Ruben Hühne

Leibniz Institute for Solid State
and Materials Research
Dresden, Germany

Seung-Hyun Moon

SuNAM Co. Ltd.
Gyeonggido, South Korea

Robert Hammond

Stanford University
Stanford, California, U.S.A.

Materials Research Society
Warrendale, Pennsylvania

CAMBRIDGE UNIVERSITY PRESS
Cambridge, New York, Melbourne, Madrid, Cape Town,
Singapore, São Paulo, Delhi, Mexico City

Cambridge University Press
32 Avenue of the Americas, New York NY 10013-2473, USA

Published in the United States of America by Cambridge University Press, New York

www.cambridge.org
Information on this title: www.cambridge.org/9781107408357

Materials Research Society
506 Keystone Drive, Warrendale, PA 15086
http://www.mrs.org

© Materials Research Society 2009

This publication has been registered with Copyright Clearance Center, Inc.
For further information please contact the Copyright Clearance Center,
Salem, Massachusetts.

First published 2009
First paperback edition 2012

Single article reprints from this publication are available through
University Microfilms Inc., 300 North Zeeb Road, Ann Arbor, MI 48106

CODEN: MRSPDH

ISBN 978-1-107-40835-7 Paperback

MILESTONES IN IBAD TEXTURING

IBAD TEXTURING

*Invited Paper

IBAD LONG LENGTH APPLICATION

*Invited Paper

TEXTURING BY OTHER TECHNIQUES

*Invited Paper

PREFACE

Thin film growth is a very old art and an established scientific field in materials science. Within this field, growth of monocrystal-like films has been practiced for many decades by making use of epitaxy on monocrystalline substrates. In the quest for greater control of materials, the next level of achievement would be to grow well-oriented thin films on arbitrary substrates, i.e., without the need of monocrystalline substrates. This is what is attempted by artificially inducing grain alignment in thin films. We use the term "artificially" to refer to film growth methods that do not utilize a monocrystalline substrate to obtain in-plane alignment. Uniaxial, out-of-plane film texture can be obtained for many materials by surface energy minimization. The absence of a template for epitaxy usually means that something else needs to determine the crystalline orientation in the plane of the film and break rotational symmetry. The source of this biaxial alignment can be an off-axis energetic beam impinging during deposition or the deposition flux itself. Over the last three decades a variety of methods for grain alignment have been demonstrated with varying degrees of success.

Symposium RR, "Artificially Induced Grain Alignment in Thin Films," held December 2–3 at the 2008 MRS Fall Meeting in Boston, Massachusetts, represents a first attempt to bring together researchers from around the world working on artificial grain alignment in films. Our particular emphasis for the symposium, based on our own experiences, was in physical vapor deposition methods for growth of inorganic thin films, with special attention paid to ion beam assisted deposition (IBAD) texturing. The symposium speakers and attendees represented a range of organizations from academia, national laboratories and industry. These proceedings capture some of about 40 presentations made at the symposium and some of the lively discussion. The symposium Round Table discussion session is transcribed starting on page 99.

We look forward to many exciting future developments in the field, and we feel confident that this field will radically change thin film growth and its applications. Grain alignment in films promises to be an enabler for a number of new technologies. We expect that a new generation of advanced film-based devices will result from the critical advancements in this field.

Vladimir Matias
Ruben Hühne
Seung-Hyun Moon
Robert Hammond

September 2009

ACKNOWLEDGMENTS

We would like to thank the following organizations for their generous support for this symposium:

Materials Research Society

Bruker HTS

k-Space Associates Inc.

Staib Instruments Inc.

Los Alamos National Laboratory

MATERIALS RESEARCH SOCIETY SYMPOSIUM PROCEEDINGS

MATERIALS RESEARCH SOCIETY SYMPOSIUM PROCEEDINGS

Prior Materials Research Society Symposium Proceed   available by contacting Materials Research Society

Milestones in IBAD Texturing

Mater. Res. Soc. Symp. Proc. Vol. 1150 © 2009 Materials Research Society   1150-RR01-01

Inducing Grain Alignment in Metals, Compounds and Multicomponent Thin Films

James M.E. Harper
Department of Physics, University of New Hampshire
Durham, NH 03824

ABSTRACT

Several methods to induce grain alignment in polycrystalline thin films are discussed, in which directional effects can dominate over the normal evolution of fiber texture during thin film growth. Early experiments with ion beam assisted deposition showed the importance of channeling directions in selecting grain orientations with low sputtering yield or low ion damage energy density. Examples of this approach include the formation of biaxial fiber textures in Nb, Al and AlN. Grain orientations may also be selected by the release of stored energy during abnormal grain growth initiated by solute precipitation (Cu-Co) or phase transformation (TiSi$_2$). Other energy sources such as mechanical deformation, crystallization or compound formation may also contribute to producing desired grain alignments. In multicomponent thin films, combinations of these mechanisms provide opportunities for more specific control of grain orientations.

INTRODUCTION

Thin film deposition often involves bombardment of the growing film by energetic particles including the depositing atoms themselves, process gas atoms (e.g. Ar) reflected from a sputtering target, ions attracted to the substrate by a bias voltage, and ion beams directed at the growing film. The latter is usually called Ion Beam Assisted Deposition (IBAD), and gives a high level of control over the deposition environment. At the request of the organizers of the Fall 2008 MRS Symposium RR on Artificially Induced Grain Alignment in Thin Films, this paper begins by summarizing the early development of IBAD at IBM Research. Next, the effects of releasing stored energy during abnormal grain growth are discussed as an approach for selecting grain orientation. Two examples are described in which this mechanism is initiated by solute precipitation and by phase transformation.

BACKGROUND

In the mid-1970's, radio frequency diode sputtering was developing as the preferred method for depositing thin films for silicon technology and memory devices. At IBM's Thomas J. Watson Research Center, Jerome J. Cuomo was studying the effects of bias sputtering on thin film properties and recognized that an ion beam could provide a more controlled bombardment environment (ion energy, flux, angle) than the plasma in a diode sputtering system. In 1976, Cuomo and the author visited several ion source manufacturers, but none provided a configuration that could be conveniently operated within an electron beam evaporator or diode sputtering system. In 1977, Cuomo initiated collaboration with Professor Harold R. Kaufman, inventor of the broad-beam, multiaperture ion source for space applications. Kaufman designed a dished set of ion source grids which enabled the beam from a 10-cm. diameter Ion Tech source to be focused onto small sputtering targets to obtain useful deposition rates. Using that source together with a 2.5-cm. diameter Ion Tech source to bombard the growing film, a dual ion beam system was set up to simulate the effects of bias sputtering on amorphous Gd-Co magnetic alloys for potential memory applications. Ion bombardment could now be quantified and the effect of preferential sputtering on composition in these materials was measured [1]. It was also observed that magnetic anisotropy in these films displayed an azimuthal uniaxial orientation that aligned with the direction of ion bombardment. Even though these films were amorphous, local anisotropy in Gd-Co pairing gave an anisotropic magnetic susceptibility. An invention disclosure was filed in April 1979, but no publication was released.

Since the 2.5-cm. diameter ion source design was too large for flexible installations, Kaufman designed a compact 2.0-cm beam diameter ion source which was installed in an electron beam evaporator at IBM Research for IBAD experiments in the summer of 1979 in collaboration with Robert H. Hammond of Stanford University [2]. Since azimuthally anisotropic properties had already been observed in GdCo alloys, we decided to look for similar effects in Nb thin films. The limited space in the vacuum chamber resulted in the ion beam being directed at 28° from normal incidence, as shown in Figure 1(a).

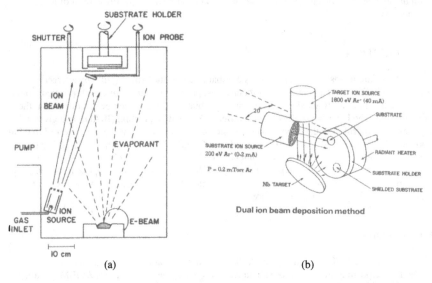

Figure 1. Configurations used for early IBAD studies of Nb texture in (a) electron beam evaporator [3] and (b) dual ion beam deposition system [4].

Niobium films evaporated at 1 nm/s on Si_3N_4 transmission electron microscopy (TEM) windows at 400 °C without ion bombardment showed no clear texture. Films deposited with 400 eV argon ions at 0.04 mA/cm^2 ion flux showed clear (110) perpendicular fiber texture, i.e. (110)

4

planes parallel to the substrate, but no azimuthal alignment. Films deposited with 800 eV argon ions at 1.1 mA/cm^2 showed clear biaxial texture with (110) planes parallel to the substrate and tilted (110) planes oriented parallel to the ion beam direction. If the ion beam had been installed closer to the (111) channeling direction at 35° from normal incidence, the biaxial texture would probably have been much stronger. No x-ray pole figure measurements were made on these films. An invention disclosure was filed in 1980 and published in 1982 with the title "Method for Controlling Crystal Orientation in Thin Films" [5].

The 2.0-cm diameter ion source was soon redesigned by Kaufman as an improved 3.0-cm diameter ion source which was the precursor of compact ion sources later manufactured by several companies including Commonwealth Scientific Corporation. More than sixty 3.0-cm ion sources were built in 1979-1981 at IBM Research and used for thin film and surface science experiments within IBM and with collaborators. A patent was issued in 1984 for the pluggable ion source design, and the 3.0-cm ion source was used for IBAD in evaporators and dual ion beam systems to explore stress modification, compound formation, tunneling layer formation and step coverage, in addition to modification of grain orientations [6]. During 1980-1982, several IBAD configurations were tested for biaxial texture in Nb, but most of the IBAD studies at IBM during that time period were focused on other properties including stress modification and compound formation, so no additional biaxial results were obtained.

In 1983, the author initiated a focused study of grain orientation in IBAD thin films by supervising an M.S. thesis project [7] by Ms. Lock See Yu, a student in the Materials Science Cooperative Studies program at the Massachusetts Institute of Technology. In this study, a dual ion beam system was configured to bombard the growing film with an Ar$^+$ ion beam at 70° from normal incidence, as shown in Figure 1(b). Without ion bombardment, strong (110) perpendicular texture was obtained at room temperature in Nb films deposited on amorphous silica substrates, as shown in the (110) pole figure in Figure 2(a). With ion bombardment, we obtained clear evidence of in-plane grain orientation (biaxial texture) in Nb thin films which grew with a strong (110) perpendicular fiber texture, as shown in Figure 2(b). The strength of the in-plane orientation was shown to increase with ion flux, as shown in Figure 2(c), although the x-ray pole figure system available could not provide quantitative information on the volume fraction of oriented grains.

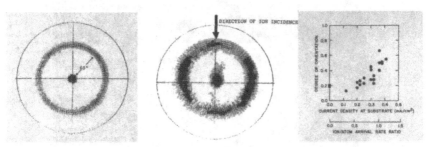

Figure 2 (a) (110) pole figure of ion beam deposited Nb film without ion bombardment, (b) with ion bombardment; (c) degree of orientation vs. ion/atom ratio [4].

5

The direction of grain orientation was shown to align with a planar (110) channeling direction, and its mirror image, in the BCC Nb crystal structure [4,8]. Promising results were also obtained on single crystal sapphire substrates. Ion bombardment during room temperature deposition produced a transition from a mixture of three orientations of (110) Nb parallel to (0001) Al_2O_3 to a single epitaxial orientation of (111) Nb parallel to (0001) Al_2O_3. An invention disclosure was submitted in April 1985, titled "Low Temperature Epitaxy and Polycrystalline Growth of Nb and Si on Sapphire", and published in 1987 [9].

These results led to a model for in-plane orientation based on the differences in sputtering yield (or other orienting mechanism) between channeling directions and non-channeling directions [10]. The model predicts that the asymptotic degree of orientation increases with ion/atom flux ratio, consistent with the Nb experimental results. Interestingly, the model also shows that the time (equivalently, the thickness) required to reach the asymptotic degree of alignment does not increase monotonically, but has a maximum as a function of the ion/atom arrival flux ratio. It is likely that this effect is responsible for variability of results on IBAD orientation in experiments that have not explored a wide range of ion/atom flux ratios. Biaxial texture in Al thin films was also demonstrated using IBAD by Srolovitz, Was et al. [11]. After the discovery of high-temperature superconductivity in 1986, IBAD was successfully used [12,13] to orient yttria-stabilized zirconia (YSZ) templates for the growth of high-temperature superconducting thin films with high critical current densities. The development of IBAD and other methods including inclined substrate deposition for template applications has not been pursued by the author of this paper, and is described by other authors in this Symposium. A model based on the energy density deposited in the growing film grains by ion impact was developed by Srolovitz [11] and shows that in-plane orientation may develop slowly with thickness, as in YSZ [12], or develop rapidly in the coalescence stage, as in MgO [14]. Additional factors that may cause changes in grain orientation include stress [15], deposition angle [16,17], orientation-dependent adatom mobility and shadowing [11,15,18].

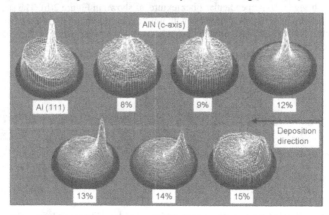

Figure 3. X-ray pole figures of Al (111) and AlN (c-axis) for $N_2/(Ar+N_2)$ gas flow ratios as indicated. The deposition direction is from the right of the figure [16].

In a study of aluminum nitride formation in dual ion beam deposition [19], the author found that AlN films grown under N_2^+ ion bombardment changed grain orientation as a function of ion energy. Also, insulating metal nitride phases such as Zr_3N_4 and Hf_3N_4 were obtained under energetic N_2^+ ion bombardment [20]. However, no detailed study of texture was made on those films. A clear demonstration of biaxial texture in AlN, measured by x-ray pole figures, was later published by Rodriguez-Navarro et al. [21] using substrate tilt to control the angle of energetic particle bombardment. Recent experiments by the author and students [16] have confirmed these observations and have shown that the AlN c-axis responds to the deposition angle (42° from normal incidence) abruptly for N_2/Ar gas flow ratios above a certain threshold, as shown in Figure 3.

These observations prompted the development of a Monte Carlo simulation model [17] that accounts for a shift in c-axis fiber texture from perpendicular to the substrate to a direction pointing towards the deposition source. In the growth of compound thin films such as AlN, the surface diffusivity of adatom species can change rapidly with a small change in reactive gas flow, changing the relative roles of lateral adatom mobility and shadowing. Additional observations on molybdenum and niobium film textures from various sputtering systems have shown the importance of deposition geometry in controlling the angles of deposition and particle bombardment, especially during the early stages of film growth that create a template for the rest of the film [18,22,23].

As film composition becomes more complex, moving from pure metals to compounds to multicomponent thin films, it is clear that more diverse mechanisms come into play in determining the eventual grain orientations in the final film. Equipment improvements have also opened up new areas for exploration. For example, R.F. ion sources without hot filaments can run on O_2 for long periods of time, allowing greater exploration of metal oxide orientations. Also, operation of magnetron deposition sources at the low end of their pressure range (1-2 mTorr) allows simultaneous operation of Kaufman ion sources at the high end of their pressure range [24]. This combination allows operation in conditions of long mean free path that maintain the directionality of both the depositing flux and the ion beam flux, which may be set at different angles to optimize orientation control. In the next section, the presence of stored energy in the film microstructure is discussed as another strong influence on grain orientations.

STORED ENERGY AND ABNORMAL GRAIN GROWTH

Thin films are metastable in many respects, and may contain stored energy which can be released to generate specific grain orientations. Here, we discuss two cases in which the release of stored energy causes strong changes in thin film texture by stimulating abnormal grain growth during annealing. While the previous discussion of IBAD effects focuses on modifying thin film properties during growth, annealing to change grain orientations is carried out after growth. The specific materials and applications will determine which approach is most feasible.

Solute precipitation

Codeposition of mutually insoluble materials may form a supersaturated solution of an alloying element within the host material. Upon annealing, the minor component may precipitate as second phase particles. At IBM Research, extensive studies of Cu alloys were carried out in the 1990's as part of developing Cu chip interconnection technology. For example, Cu and Co have very limited solubility in equilibrium. Cu alloys containing several atomic percent Co have high resistivity following deposition, since the Co atoms are dispersed in a metastable solution, as shown in Figure 4(a) [25].

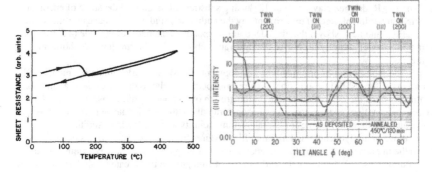

Figure 4 (a) Sheet resistance vs. temperature for Cu-Co alloy (b) (111) pole figure of Cu-Co alloy before and after annealing to 450 °C for 120 min [25]

Upon annealing to a temperature of 100-250 °C, depending on the Co content, the resistivity decreases abruptly as the Co is precipitated from solution. Simultaneously, the film stress is reduced. This precipitation process forms Co particles about 10 nm in diameter, and also causes a dramatic change in film texture. These results were found for both coevaporated and electroplated Cu-Co alloys. As shown in Figure 4(b) an as-deposited electroplated Cu-0.44 at. % Co alloy has a moderately strong (111) texture perpendicular to the substrate surface, with weaker (200) components and twin orientations. After annealing to 450 °C for 120 min, the texture is strongly (200) perpendicular to the substrate surface, resulting from abnormal grain growth of grains with this orientation. A TEM image of a large grain in an annealed electroplated Cu-0.7 at. % Co film is shown in Figure 5, with orthogonal twin boundaries typical of Cu.

Figure 5. TEM plan view image of thinned 1 μm thick Cu-Co alloy film after annealing [25].

8

The diameter of these (200) oriented grains is in the range of 20 times the film thickness. Coupling of solute precipitation to abnormal grain growth is also called discontinuous precipitation [26], highlighting the fact that most of the precipitation of solute atoms into solute particles takes place abruptly along the moving boundary of the abnormally growing grains.

The stored energy in the Cu-Co alloys that generates highly oriented grains is released by Co solute precipitation. Both the energy of solution and the grain boundary energy density are reduced by forming Co particles at the same time as forming much larger grains. For comparison, in pure Cu, abnormal grain growth is typically observed only for annealing temperatures above 500 °C [27]. The mechanism that keeps the grain boundaries moving in Cu-Co alloys is the additional release of energy caused by solute atom precipitation. Estimates indicate that the energy density available from solute precipitation can easily exceed the grain boundary energy density [28]. In developing methods for artificial control of grain orientations, we have an opportunity to take advantage of the inherent tendency for grain reorientation provided by this internal energy source. This mechanism also draws attention to materials that might be successfully alloyed with a small fraction of second-phase material in order to introduce solute precipitation as an energy source, without compromising the desired properties of the host material.

Phase transformation

Many useful thin film materials have structural phase transformations which must be controlled to obtain desired properties. For example, the compound $TiSi_2$, used in silicon device contacts, is formed by thermal reaction of Ti with Si. The phase initially formed is the C49 structure, which has a high resistivity (50-75 $\mu\Omega$-cm) and small grain size (typically tens of nm) [29]. The desired phase of $TiSi_2$ (resistivity 15-20 $\mu\Omega$-cm) has the C54 structure, which is usually obtained by a high temperature anneal around 800 °C. At IBM Research, a substantial effort was made in the 1990's to extend the use of $TiSi_2$ to the smallest possible feature sizes [30] before it became necessary to use $CoSi_2$. As part of these studies, the phase transformation was examined in great detail, including its effect on grain orientations in small feature sizes. The author proposed a study of grain orientations as an M.S. thesis topic for Vjekoslav Svilan, a student in the Electrical Engineering Cooperative Studies Program at the Massachusetts Institute of Technology. His thesis, "Texture Analysis in Submicron Structures of Titanium Silicide" was completed in 1996 and the key results were published in 1997 [31]. Patterned lines of polycrystalline Si were reacted with Ti to form C49 $TiSi_2$, followed by annealing to form C54 $TiSi_2$. Narrow lines of 0.22 μm width showed a strong orientation of the C54 grains along the direction of the narrow lines. While the C49 grains were smaller than the line width, the C54 grains extended along the lines to distances about 15 times the line width. Given the small amount of material present in these patterned thin layers, the X20C beam line at the National Synchrotron Light Source was used to provide a very high x-ray flux which was able to follow the phase transformation in real time and obtain accurate pole figures. The C54 grains were found to be strongly textured with the (040) direction perpendicular to the substrate surface, and the (100) direction aligned with the narrow line direction, as shown in Figure 6. In this non-cubic structure, diffraction from the (100) planes could not be measured directly, since that scattering vector was in the plane of the sample. Instead, pole figures of the (040), (311) and (022) planes were examined to precisely determine the (100) direction as being parallel to the patterned line direction. A sketch of the orientation of C54 grains with respect to the lines is shown in Figure 7(a) and examples of long C54 grains extending along the narrow lines are given in Figure 7(b).

These results provide an example of abnormal grain growth stimulated by a structural phase transformation. The grains that grow fastest along the line direction are the grains that dominate the final population, therefore anisotropic grain growth velocity is an important parameter in selecting the grain orientation in confined structures. The resulting large grain orientation is determined by the physical constraints of the patterned film instead of an externally imposed ion bombardment direction.

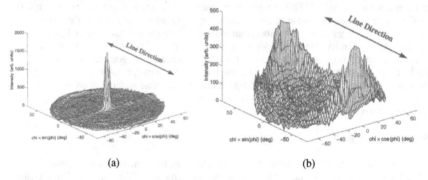

(a)   (b)

Figure 6 (a) (040) pole figure and (b) (022) pole figure of C54 TiSi$_2$ patterned lines [31].

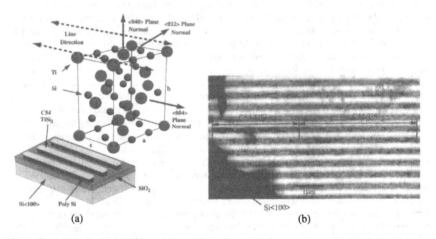

(a)   (b)

Figure 7 (a) Sketch of orientation of C54 TiSi$_2$ unit cell relative to patterned polycrystalline Si lines, (b) TEM plan view image of C54 grains extending along the lines, indicated by arrows [31].

DISCUSSION

Inducing grain orientations by IBAD typically requires depositing an energy density corresponding to 1-10 eV per atom in the form of ion damage into the growing film [28]. An exception is when a template layer is well oriented by IBAD and subsequent growth maintains the orientation by granular epitaxy, in which case the ion flux can be decreased or shut off. By taking advantage of energy densities contained in the film itself, we have a source of grain reorientation that can be manipulated. In the example of Cu-Co solute precipitation and abnormal grain growth, the azimuthal orientation of the large (200) textured grains is presumably determined by the orientation of the first grains that start to grow rapidly during annealing. These grains are selected by variables including local variations in solute concentration, grain boundary misorientations and grain boundary pinning by impurities. In fact, grain boundary velocity can have a nonmonotonic dependence on driving force for certain solute concentrations [32], allowing the possibility of an abrupt increase in velocity as certain grains overcome the pinning effect of the solute atoms. In order to impose a desired azimuthal orientation on the large oriented grains, we must introduce a directional influence that selects grains to start growing rapidly with the desired orientation. Here is an opportunity for ion bombardment to be applied during annealing, so that the grain orientations that start to grow at a critical temperature are those favored by the selection mechanisms inherent in ion bombardment. Ideally, a brief exposure to ion bombardment at a specific angle will stimulate abnormal grain growth with a chosen orientation. Once these grains start to grow, the ion beam orientation mechanism should no longer be needed, and the beam can be turned off.

Similarly, in cases where a phase transformation initiates abnormal grain growth, as in the C49-to-C54 $TiSi_2$ phase transformation, the orientation of newly growing grains may be determined by physical patterning, as in the above example, or could be determined by the application of an orienting ion beam. In other work at IBM, the C54 $TiSi_2$ formation temperature was successfully lowered, and the nucleation site density for the transformation increased, by adding several atomic percent of specific transition metals such as Ta and Nb [30]. Whether these additives also changed the grain orientation in C54 $TiSi_2$ was not determined, but would be an interesting question to answer.

Other internal energy sources include mechanical deformation, which has been suggested as a driving force for abnormal grain growth in electroplated Cu thin films [33], internal stress [15], crystallization from an amorphous precursor state [34], and compound formation within the film or with a multilayer film structure. The energy densities for some examples of these processes have been estimated [28] and are also found to exceed the energy density driving normal grain growth.

CONCLUSIONS

The early development of IBAD methods depended strongly on new designs of compact ion beam sources that could be installed in deposition chambers mainly used for other purposes. The 3.0-cm Kaufman ion source, in particular, was used in numerous experiments in thin film modification that provided quantitative information on beam flux, energy and angle that was unavailable in plasma deposition systems. Grain orientations obtained in IBAD were generally consistent with selection of channeling directions by the ion beam. However, these methods typically require depositing ion energy in the range of 1-10 eV per atom, which damages and resputters much of the growing film. Here, it is suggested to explore the energy density of other mechanisms to induce desired grain orientations via abnormal grain growth. Two examples are described, namely solute precipitation and phase transformation. Other sources of stored energy including mechanical deformation, internal stress, crystallization and compound formation, are potentially powerful mechanisms for changing grain orientations. Multicomponent thin films offer combinations of these mechanisms that might be successful in creating artificially induced grain orientations for specific applications.

ACKNOWLEDGMENTS

The author acknowledges many colleagues who influenced the development of IBAD, especially for grain orientation; at IBM Research, Jerome J. Cuomo, Richard J. Gambino, Sigrid Herd, David A. Smith, Kenneth P. Rodbell, Christian Lavoie and Cyril Cabral, Jr.; at Colorado State University, Professors Harold Kaufman and Mark Bradley; at Stanford University, Robert H. Hammond and Connie Wang; at MIT, Lock-See Yu and Vjekoslav Svilan. Also David Srolovitz, Steven Yalisove, Christian Weissmantel, Joe Greene, David Hoffman and Guy Gautherin. He also acknowledges the more recent help of students at the University of New Hampshire, Derya Deniz, Dana Filoti, Amanda Brown, Anne Marie Shover, Nicholas Dellas, Katherine Hamilton and Don Carlson, as well as Professors James Krzanowski, Bryan Huey, Tansel Karabacak, Daniel Gall and Xiaoman Duan.

REFERENCES

1. J.M.E. Harper and R.J. Gambino, J. Vac. Sci. Technol. Vol. 16, 1901 (1979).
2. Private Communication - letter of 8/16/95 from Harper to R.H. Hammond summarizing all IBAD experiments on biaxial texture at IBM 1978-1985.
3. J.J. Cuomo, J.M.E. Harper, C.R. Guarnieri, D.S. Yee, L.J. Attanasio, J. Angilello, C.T. Wu and R.H. Hammond, J. Vac. Sci. Technol. Vol. 20, 349 (1982).
4. L.S. Yu, J.M.E. Harper, J.J. Cuomo and D.A. Smith, Appl. Phys. Lett. 47, 932 (1985).
5. J.J. Cuomo, C.R. Guarnieri, R.H. Hammond, J.M.E. Harper, S. Herd, D.S. Yu, IBM Invention Disclosure 2/11/80, IBM Tech. Discl. Bull. 25, 3331 (1982).
6. J.J. Cuomo, J.M.E. Harper, H.R. Kaufman, U.S. Patent No. 4446403 (1 May 1984); H.R. Kaufman, J.J. Cuomo and J.M.E. Harper, J. Vac. Sci. Technol. Vol. 21, 725 (1982).
7. *"The Effect of Low Energy Ion Bombardment on the Crystallographic Orientation of Thin Films"*, Lock See Yu, M.S. Thesis, Massachusetts Institute of Technology, September 1985.
8. J.M.E. Harper, D.A. Smith, L.S. Yu and J.J. Cuomo, Proc. MRS 51, 343 (1986).
9. J.J. Cuomo, J.M.E. Harper and L.S. Yu, IBM Tech. Discl. Bull. 29, 4492 (1987).
10. R.M. Bradley, J.M.E. Harper and D.A. Smith, J. Appl. Phys. 60, 4160 (1986).
11. L. Dong, D.J. Srolovitz, G.S. Was, Q. Zhao and A.D. Rollett, J. Mater. Res. 16, 210 (2001).
12. Y. Iijima, N. Tanabe, O. Kohno and Y. Ikeno, Appl. Phys. Lett. 60, 769 (1992).

13. P.N. Arendt, S.R. Foltyn, J.R. Groves, R.F. DePaula, P.C. Dowden, J.M. Roper and J.Y. Coulter, Appl. Supercond. 4, 429 (1998).
14. C.P. Wang, K.B. Do, M.R. Beasley, T.H. Geballe and R.H. Hammond, Appl. Phys. Lett. 71, 2955 (1997).
15. S. Mahieu, P.Ghekiere, D. Depla and R. De Gryse, Thin Solid Films 515, 1229 (2006).
16. D. Deniz, J.M.E. Harper, J.W. Hoehn and F. Chen, J. Vac. Sci. Technol. A25, 1214 (2007).
17. D. Deniz, T. Karabacak and J.M.E. Harper, J. Appl. Phys. 103, 83553 (2008).
18. O. Karpenko, J.C. Bilello and S.M. Yalisove, J. Appl. Phys. 76, 4610 (1994).
19. J.M.E Harper, J.J. Cuomo and H.T.G. Hentzell, J. Appl. Phys. 58, 550 (1985).
20. B.O. Johansson, H.T.G. Hentzell, J.M.E. Harper and J.J. Cuomo, J. Mat. Res. 1, 442 (1986).
21. A. Rodriguez-Navarro, W. Otano-Rivera, J. M. Garcia-Ruiz, R. Messier and L. J. Pilione, J. Mat. Res. 12, 1689 (1997).
22. J.M.E. Harper, K.P. Rodbell, E.G. Colgan and R.H. Hammond, J. Appl. Phys. 82, 4319 (1997).
23. G.S. Was, J.W. Jones, C.E. Kalnas, L.J. Parfitt, A. Mashayekhi and D.W. Hoffman, Nucl. Instr. and Methods in Physics B80/81 1356 (1993).
24. N.S. Dellas and J.M.E. Harper, Thin Solid Films 515, 1646 (2006).
25. J.M.E. Harper, J. Gupta, D.A. Smith, J.W. Chang, K.L. Holloway, C. Cabral Jr., D.P. Tracy and D.B. Knorr, Appl. Phys. Lett. 65, 177 (1994).
26. *Recrystallization of Metallic Materials*, F. Haessner, ed. (Dr. Riederer Verlag GmbH, Stuttgart, 1978), p. 140.
27. J. Gupta, J.M.E. Harper, J.L. Mauer IV, P.G. Blauner and D.A. Smith, Appl. Phys. Lett. 61, 663 (1992).
28. J.M.E. Harper and K.P. Rodbell, J. Vac. Sci. Technol. B15, 763 (1997).
29. J.M.E. Harper, C. Cabral Jr. and C. Lavoie, Ann. Rev. Mat. Sci. 30, 523 (2000).
30. R.W. Mann, G.L. Miles, T.A. Knotts, D.W. Rakowski, L.A. Clevenger, J.M.E. Harper, F.M. d'Heurle and C. Cabral Jr., Appl. Phys. Lett. 67, 3729 (1995).
31. V. Svilan, K.P. Rodbell, L.A. Clevenger, C. Cabral Jr., R.A. Roy, I.C. Noyan, C. Lavoie, J. Jordan-Sweet and J.M.E. Harper, J. Elec. Mat. 26, 1090 (1997).
32. K. Lücke and H.P. Stüwe, in *Recovery and Recrystallization of Metals,* L. Himmel, ed. (Interscience, New York, 1963), p. 160.
33. C. Detavernier, D. Deduytsche, R.L. Van Meirhaeghe, J. De Baerdemaeker and C. Dauwe, Appl. Phys. Lett. 82, 1863 (2003).
34. K. T-Y. Kung, R.B. Iverson and R. Reif, Appl. Phys. Lett. 46, 683 (1985).

Mater. Res. Soc. Symp. Proc. Vol. 1150 © 2009 Materials Research Society   1150-RR01-02

Development of IBAD Process for Biaxial Texture Control of RE-123 Coated Conductors

Yasuhiro Iijima[1]
[1]Fujikura Ltd., 1440 Mutsuzaki, Sakura-shi, Chiba 285-8550, Japan

ABSTRACT

Ion-Beam-Assisted Deposition (IBAD) is a sophisticated technique to deposit biaxially textured thin films directly on non-textured substrates by using concurrent off-normal ion beam bombardment. It is a key technology for $REBa_2Cu_3O_{7-x}$ (RE-123) coated conductor, which is the superconducting wire with the largest I_c performance ever available, favorable for various practical applications operating around liquid nitrogen temperature. This paper reviews development of materials and vacuum technology for IBAD technique, which contributed to the drastic advancement of long length, high-performance, and cost effective RE-123 coated conductor, achieved in this two decades.

INTRODUCTION

When cuprate superconductors with T_c over nitrogen boiling point were discovered in late 1980s, tremendous research efforts started to develop them into flexible wires. Above all, RE-123 superconducting materials have the largest J_c performance, but it has so sensitive intergranular weaklinks that easily interrupt macroscopic transporting current. RE-123 films epitaxially grown on single crystal substrates by vapor phase coating techniques had revealed excellent J_c performance in early works, but it was very difficult to form "biaxially aligned structure" (single-crystal-like structure without large-angle grain boundaries) in long flexible polycrystalline conductors, that is the only solution to get sufficient transport current of RE-123 conductors by eliminating weaklinks come from grain boundaries with misalignment angles of even a few degrees[1].

In 1991, the first high-J_c Y-123 film was grown on a polished metal tape with a biaxially textured template layer deposited by off-normal IBAD technique[2-3]. Sharply in-plane textured fine oxide surface formed on flexible non-textured metal tapes as Ni-Cr alloy or stainless steel is the ideal substrate structure for RE-123 conductors with both good mechanical strength, and flat, hard, and chemically stable surface. So far the best performance were constantly reported for RE-123 conductors made by using IBAD approach, than the major counterpart technique using metallurgically textured metal tapes of Ni, etc[4]. This paper describes the development of IBAD template buffer layer materials and reel-to-reel continuous processing with large-scale ion source technology, both of which have dramatically improved in these two decades.

Fluorite type oxides

The in-plane ordering of thin films by concurrent off-normal ion bombardment during growth was first reported in Nb metal films on fused silica in 1985[5]. Sharp cube-textured structure was obtained at 1991, when strong out-of-plane <100> alignment was found in yttria-stabilized-zirconia (YSZ) films grown on polished Ni-Cr alloy substrate by IBAD. An <100> axis of YSZ strongly aligns normal during deposition with concurrent low-energy (~200eV) ion bombardment, at the temperature below 300°C. The ion incident angle was optimized 55 degrees inclined from substrate normal, which corresponded to an <111> axis. Figure 1 shows TEM cross-sectional image of a YSZ film. At the initial growth stage, the structure was just non-textured nano-crystallites. <100> aligned fiber texture was gradually evolved during growth[6]. Figure 2 shows thickness dependence of in-plane mosaic spread for YSZ and $Gd_2Zr_2O_7$. The results agreed with crystalline orientation model that should enhance during growth by collaboration of homoepitaxy and selective growth come from anisotropic resputtering and radiation damage determined by ion-channeling.

Table I shows deposition conditions and optimized mosaic spreads for varied $(ZrO_2)_{2x}$-$(RE_2O_3)_{1-x}$ films of ~1μm thick. In these films crystalline structure change from fluorite to rare earth C. They have basically similar structure but mean valence number of cation change from 4+ to 3+. Lattice binding energy density is calculated from Born-Habar cycle and lattice constant. A trend was observed that the better textured growth was obtained for large lattice energy density, which suggest excess radiation damage deteriorate homo-epitaxial growth. It is noteworthy that pyrochlore composition oxides (Zr:RE=1:1) had the least mosaic spread and shorter time constant for texture evolution as shown in figure 2. It suggests gain of cation mixing free energy etc. should contribute to the durability for radiation damage[7-9].

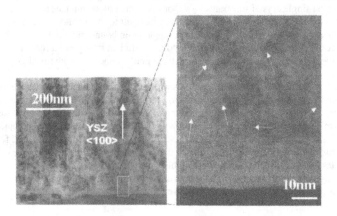

Figure 1. TEM cross-sectional image of a YSZ film deposited on polished Ni-Cr alloy by IBAD. Amorphous and non-textured YSZ crystallites were observed at the initial growth stage.

Fluorite type IBAD oxide template films were widely used for examination of RE-123 coated conductors. But it had a problem of low throughput and high cost for the requirement of thickness around 1μm, come from growth evolution structure described above.

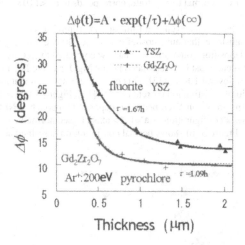

$$\Delta\phi(t)=A \cdot \exp(t/\tau)+\Delta\phi(\infty)$$

Figure 2. Thickness dependence of in-plane mosaic spread for YSZ and Gd$_2$Zr$_2$O$_7$. The results agreed with crystalline orientation model proposed by Bradley, et.al.[10].

Table I. Optimized deposition parameters and typical Δφ values for various ZrO$_2$-RE$_2$O$_3$ films.

Composition	Temperature (°C)	Ar$^+$ ion Energy (eV)	Typical Δφ values*(deg.)	Lattice energy density** (eV/nm³)	Lattice const.(nm)
ZrO$_2$-Y$_2$O$_3$ (92:8)	R.T.	200	15.4	3192	0.514
ZrO$_2$-Gd$_2$O$_3$ (92:8)	150	200	15.8	-	-
Yb$_2$Zr$_2$O$_7$	200	200	15.5	2703	0.517
Y$_2$Zr$_2$O$_7$	200	200	12.3	2643	1.038
Gd$_2$Zr$_2$O$_7$	200	200	11.3	2532	1.052
Sm$_2$Zr$_2$O$_7$	200	200	12.2	2472	1.059
Yb$_2$O$_3$	200	200	17.3	2023	1.043
Y$_2$O$_3$	300	150	24.5	1873	1.060
Gd$_2$O$_3$	-	-	random	1753	1.081
Sm$_2$O$_3$	-	-	random	1677	1.093

*film thickness near 1.0 μm.
**estimated by using the data for ZrO$_2$ and RE$_2$O$_3$ derived from Born-Habar cycle.

Halite type oxides

In 1996, Stanford Univ. group found new series of IBAD-template materials of halite type, MgO, etc[11]. Halite type nitride films of TiN etc. also found[12]. Optimized ion incident angle is 45 degrees inclined from normal to substrate, corresponds to an <110> axis where out-of-plane alignment is <100>.

In contrast to fluorite type films, even thin films around 10nm had very sharp in-plane alignment on non-textured substrate, that quite cost effective process could be expected. On the other hand, the window of deposition conditions was narrow compared to fluorite ones. Suitable bed layer of amorphous Y_2O_3 etc. must be intercalated below MgO, and the favorable MgO thickness were limited below several 10s nm[13]. This paper presents alignment axis transition determined by ion/atom ratio or assisting ion current density [14]. Figure 3 shows two kinds of alignment axes for MgO films grown on Y_2O_3 bed layer deposited on polished Ni-Cr alloy substrates. Figure 3 (a) shows a configuration of a <111> axis aligned normal, and an <100> axis aligned to beam axis, where Figure 3 (b) shows normal configuration described above.

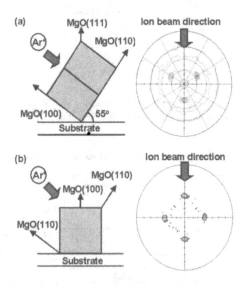

Figure 3. (a) crystalline alignment axes for MgO with an <111> axis aligned normal; (b) crystalline alignment axis for MgO with an <100> axis aligned normal. MgO (110) poles were measured by XRD.

Figure 4 shows the relationship between ion current density and in-plane textures of the IBAD-MgO films. Shapes of the symbol correspond to the in-plane texture of IBAD-MgO films. Deposition rates were 0.23 nm/sec, 0.26 nm/sec and 0.29 nm/sec, and incident angle of the ion beam were 55°, 45° and 45° respectively. For all deposition rates, in-plane textures of the <111> aligned IBAD-MgO films became sharper with increasing ion current densities. Some minimum points were observed around 90-100 μA/cm² for 0.26 nm/sec and 0.29 nm/sec. At further ion current density, the texture of the MgO change drastically. Specifically <100> aligned MgO films appeared instead of the <111> aligned MgO. The ion/atom ratio at the transition point is 0.2-0.3.

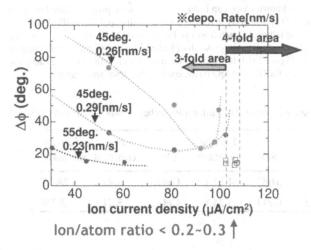

Figure 4. Ion current density dependence for in-plane textures of the IBAD-MgO films. Beam energy was set 800-900eV. The square symbols refer to the <100> 4-fold samples.

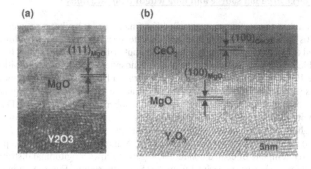

Figure 5. HRTEM (high-resolution TEM) images at the (a) MgO/Y₂O₃ interface for <111> aligned MgO, and (b) MgO/Y₂O₃ interface and CeO₂/MgO interface for <100> aligned MgO.

Figure 5 shows the HRTEM (high-resolution TEM) images at the (a) MgO/Y_2O_3 interface for <111> aligned MgO, and (b) MgO/Y_2O_3 interface for <100> aligned MgO. It was shown that MgO(111) is also directly on the Y_2O_3, similar to MgO(100) case. Sometimes it is said that MgO(100) or TiN(100) tilt to the ion beam at a thick films[12-13]. But this was thought to be different cases, which determined in nucleation stage of MgO films. It is clearly contrasted from the initial growth stage of IBAD-fluorite films.

It is well known CeO_2 cap layer is effective to improve surface texture of fluorite-type IBAD-template [15]. Recently it was also successfully applied on IBAD-MgO template films [16]. Though the lattice mismatch was so big between MgO and CeO_2 as shown in Table II, CeO_2 (100) plane was directly grown by PLD on MgO (100) plane with cube-on-cube symmetry, as indicated in Figure 5 (b). It means lattice mismatch relaxation using $LaMnO_3$ etc. are not always necessary for IBAD-MgO based coated conductor architecture.

Table II. Lattice constants and lattice mismatches of MgO, GZO and CeO_2.

	sub.	Lattice constant	film	Lattice constant	Lattice mismatch
1	MgO	4.21Å	CeO_2	5.41Å	+28.5%
2	GZO	5.24Å	CeO_2	5.41Å	+3.2%

DEVELOPMENT OF REEL-TO-REEL IBAD PROCESSING

Development of large area ion source and long length template films

Broad beam, low energy ion sources called Kaufman type, or bucket type, were developed in 1970s for ion propulsion of artificial satellite, or nuclear fusion technology, etc. They were applied for vacuum processing as ion beam sputtering, etc. in 1980s [17] but there were not so reliable ion source that operate in oxygen atmosphere required for deposition of oxide films. Late 1990s, the situation changed by the demand from optoelectronics industry. Figure 6(a) shows a photo of 6x66cm RF discharged linear ion source, which has a guarantee of stable operation over 500 hours even in pure oxygen atmosphere. A reel-to-reel IBAD apparatus was constructed with the ion sources and 100-200 m long good textured $Gd_2Zr_2O_7$ films were grown with throughput of 1 m/h, one of which was used in the first 100m long RE-123 wire[18].

Figure 6(b) shows the world largest ion source with diameter of 15x110cm, especially designed for textured template of coated conductor. Figure 7 shows longitudinal distributions of ion current density for an Ar+ beam derived from the15x110cm source. Figure 8 shows a photo and schematic diagram of the world largest reel-to-reel IBAD system. 500-m long $Gd_2Zr_2O_7$ films were formed with throughput of 5-7 m/hour[19].

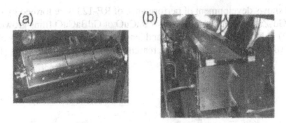

Figure 6. Photos for RF discharged linear ion sources; (a) 6x66cm type. (b) 15x110cm type, especially designed for RE-123 coated conductor processing.

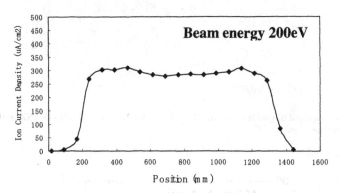

Figure 7. Longitudinal distributions of ion current density for an Ar+ beam derived from the15x110cm source. Beam transmission length was ~30cm.

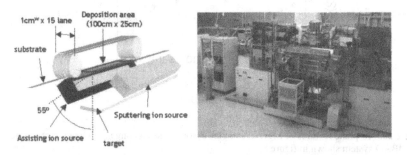

Figure 8. A photo and schematic diagram of the large scale reel-to-reel IBAD system, equipped with the15x110cm source.

Figure 9 shows development of performance of RE-123 wire formed by using IBAD-YSZ or IBAD- $Gd_2Zr_2O_7$ template films and YBaCuO or GdBaCuO films grown by pulsed laser deposition (PLD) in Fujikura. Both I_c value and length (L) dramatically improved with the development of IBAD-apparatuses. I_c x L increased 10 times every three years [19].

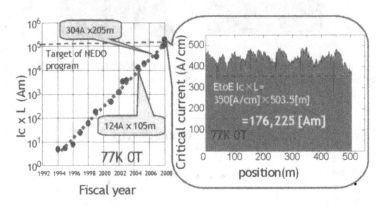

Figure 9. Development of performance of RE-123 wire formed by using IBAD-YSZ or IBAD-$Gd_2Zr_2O_7$ template films.

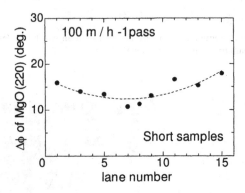

Figure 10. In-plane textures of MgO of samples at some lanes among all 15 lanes of the large scale IBAD system shown in figure 8.

Improvement of processing throughput by using MgO template

In the years around 2005, Los Alamos National Lab. and SuperPower company achieved big progress for IBAD-MgO based coated conductor technology [20]. This paper describes preliminary results of high-throughput IBAD-MgO processing by using 15cm x 110cm ion source. Figure 10 shows in-plane textures of MgO of samples at some lanes among all 15 lanes of the IBAD system shown in figure 8. The samples fabricated at 100 m/h – 1pass. MgO films with in-plane alignment were fabricated at very wide area [19].

Then we fabricated several 100 m-length IBAD-MgO films by using center 2 lanes with throughput of 100m/h, and by using center 5 lanes with throughput of 500m/h. In-plane textures of those MgO templates were $\Delta\Phi= 9\sim11°$. $\Delta\Phi= 4\sim5°$ were obtained for PLD-CeO$_2$ films grown on IBAD-MgO substrates. Table III shows I_c/J_c results for GdBaCuO films deposited by PLD on these substrates. I_c over 500 A was obtained in a 1cm wide short sample, and $I_c \sim 350$ A was obtained for a 30-m long sample using IBAD-MgO formed with throughput of 500m/h [16].

Table III. I_c/J_c results for GdBaCuO films deposited by PLD on IBAD-MgO templates formed by the large scale IBAD system shown in figure 8.

Lane No. used in IBAD	Tape speed for IBAD-MgO processing (m/h)	Length (m)	I_c (A/cm)	J_c (A/cm^2)
2	100	short	360	3.7M
2	100	short	550	2.7M
2	100	~10	400	~2M
2	100	~10	300	~2.5M
5	500	~30	350	~3.5M

CONCLUSIONS

Biaxially aligned fluorite and halite type oxide films were found to grow on non-textured substrate by off-normal IBAD. Both types were successfully applied for high –performance RE-123 coated conductors. Growth structure is very different between the two type films. Though the mosaic spreads of fluorite type films are determined by gradual texture evolution during growth, the ones for halite films are fixed at the initial nucleation stage.

By recent ion source technology development , reel-to-reel coating processing is quite reliable and promising approach for IBAD-textured template. Furthermore, thin halite type oxide film has attained large throughput of 500m/h for 1cm wide tape. IBAD is now the most promising approach for high-performance, and cost effective RE-123 coated conductors.

ACKNOWLEDGMENTS

This paper includes the experimental results obtained in programs supported by the new Energy and Industrial Technology Development Organization (NEDO).

REFERENCES

1. D. Dimos, P. Chaudhari, and J. Mannhart, *Phys.Rev.* **B41** 4038(1990).
2. Y. Iijima, N. Tanabe, Y. Ikeno, O. Kohno, *Physica C* **185-189** 1959(1991).
3. Y.Iijima, N.Tanabe, O.Kohno and Y.Ikeno, *Appl. Phys. Lett.* **60** 769(1992).
4. Y.Iijima and K.Matsumoto, *Supercond. Sci. & Tech.* **13** (1) 68(2000).
5. L.S.Yu, J.M.E.Harper, J.J.Cuomo, and D.A.Smith, *Appl.Phys.Lett.* **47** 932 (1985).
6. Y.Iijima, M.Hosaka, N.Tanabe, N.Sadakata, T.Saitoh, O.Kohno and K.Takeda, *J. Mat. Res.* **13** (11) 3106(1998).
7. Y. Iijima, M. Kimura, and T. Saitoh: in *Fundamental mechanisms of low-energy ion beam induced surface growth* (Mat. Rres. Soc. Symp. Proc. **585**, Warrendale , PA, 2000) p45.
8. Y. Iijima, K. Kakimoto, Y. Yamada, T. Izumi, T. Saitoh and Y. Shiohara: *MRS Bull.* **29**, No.8, 564(2004).
9. Y. Iijima, K. Kakimoto, T. Saitoh, T. Katoh, and T. Hirayama, *Physica C* **378-381**, Part2 960(2002).
10. R.M.Bradley, J.M.E.Harper, and D.A.Smith, *J.Appl.Phys.* **60** 4160(1986).
11. C.P.Wang, K.B.Do, M.R.Beasley, T.H.Geballe, and R.H.Hammond, *Appl. Phys.Lett.* **71** 2955(1997).
12. R. Hühne, S. Fähler, L. Schultz, and B. Holzapfel, *Physica C* **426-431**, 893(2005).
13. P. N. Arendt, S. R. Foltyn, *MRS Bull.* **29** 543(2004).
14. S. Hanyu, T. Miura, Y. Iijima, M. Igarashi, Y. Hanada, H. Fuji, K. Kakimoto, T. Kato, T. Hirayama and T. Saitoh, *Physica C* **468**, 1561(2008).
15. T.Muroga, T.Araki, T.Niwa, Y.Iijima, T.Saitoh, I.Hirabayashi, Y.Yamada, and Y.Shiohara, *IEEE Trans. Appl. Supercond.* **13** 2532(2003).
16. S. Hanyu et.al., submitted to *Physica C.*
17. J. J. Cuomo, S. M. Rossnagel, and H. R. Kaufman: *Handbook of Ion-Beam Processing Technology*, p.170, Noyes, New Jersey, 1989
18. Y. Iijima, K. Kakimoto, Y. Sutoh, S. Ajimura and T. Saitoh, *IEEE Trans. Appl. Supercond.* **15** 2590(2005).
19. S. Hanyu et.al., submitted to *IEEE Trans. Appl. Supercond.*
20. V. Selvamanickam, Y. Chen, X. Xiong, Y. Y. Xie, J. L. Reeves, X. Zhang, Y. Qiao, K. P. Lenseth, R. M. Schmidt, A. Rar, D. W. Hazelton, K. Tekletsadik, *IEEE Trans. Appl. Supercond.* **17** No.2, 3231 (2007).

Mater. Res. Soc. Symp. Proc. Vol. 1150 © 2009 Materials Research Society   1150-RR01-03

From IBM (Harper) to Stanford, from IBAD-YSZ to ITaN-MgO

Robert H. Hammond
Geballe Laboratory for Advanced Materials
Stanford University
Stanford, Calif.

ABSTRACT

This session is titled "Milestones in IBAD Texturing". Three talks in this session contain history and status of the development of IBAD for texture, starting with James Harper, followed by Yasuhiro Iijima on the history and status of IBAD-YSZ first used for HTSC Coated Conductor. This paper continues the history including the discovery of IBAD-MgO. Subsequent developments in the understanding of the mechanisms of bi-axial texture through experiments and theory are reviewed to arrive at the present but not complete understanding. New in-situ characterization needs are discussed and new tools to affect the texture development are suggested.

INTRODUCTION

The title of this session is " Milestones in IBAD Texturing". The first talk and paper is Jim Harper's review and history of the early results on "Inducing Grain Alignment in Metals, Compounds and Multicomponent Thin Films", including up to his recent results[1]. It was due to his (and others at IBM-Yorktown) efforts that Harold Kaufman developed the compact ion beam source that has made ion texturing possible. This came to fruition in 1979, in time for the summer visit of the author to learn about ion beams and their effect on thin films.

The second talk and paper covers the important development of IBAD-YSZ for the High Temperature Superconductors application known as Coated Conductors. This was originated by Dr. Iijima at Fujikura Ltd., in the early 1990's[2]. The success of this development showed that this could be a route to the successful answer to the demanding requirements of the ceramic materials requiring essentially kilometers of biaxial texturing on top of flexible metal tapes.

The third talk and this paper are in part the description of the discovery and development of a more economical method, IBAD – MgO, using ion texturing at nucleation, or ITaN[3]. The idea for this originated while at IBM the summer of 1979, while discussing observations of texturing during certain sputtering operations with Harper. The vision was to affect the nucleation of atoms as they nucleated using ions, and to observe the effect with RHEED (Refection High Energy Electron Diffraction) in real time. It is this in-situ characterization that distinguishes this development. Biaxial texturing was being observed after sputtering and ion beam IBAD, but there was no idea of when this was occurring, and usually found only in thick films approaching a micron in thickness. That mechanism is by evolution, as described by Bradley et al in 1986[4].

The current research results at Stanford are in the papers by James Groves.[5, 6]

Early Efforts at Stanford

Motivated by the success with YBCO on IBAD-YSZ by Iijima[7], Paul Arendt (LANL)[8, 9] and Reade and Berdahl (LBL)[10], an effort was initiated at Stanford University under the sponsorship of Ted Geballe and funded by Paul Gant at EPRI and made possible by the new Molecular Beam Synthesis (MBS) facility (funded by a proposal of Mac Beasley) that was designed with the ITaN requirements in mind: a RHEED system, Kaufman ion source, evaporation sources, rate control, and 4-axis sample orientation. Graduate student Khiem Do initiated the work using MgO because it has a (100) preferred growth and thus did not require overcoming another texture (as is the case with YSZ), and because its lattice constant is not too bad for YBCO.

The procedure used involved fixing the ion flux and increasing the MgO flux in steps, pausing 20 seconds before increasing it again, watching the RHEED for a pattern different from the amorphous pattern of the SiN bed layer. A RHEED pattern did appear at a certain flux, on the very first attempt. The ratio of ion to MgO flux at this point is called the critical ratio. For later reference note that until the critical ratio is reached in this mode there is no deposition of MgO—it is all sputtered off by the ion beam. This we call the Stanford mode. It was found that the RHEED pattern improved as the deposition continued at the same ratio, until saturation was reached. This occurred at a thickness of between 60 and 100 Å. In order to measure the texture quantitatively by XRD, a thick MgO layer was deposited homoepitaxially at some higher temperature, say 400 °C. Connie Wang continued this research, exploring many factors to find the best conditions[11, 12]. The factors that were found to be important included: temperature, ion/molecule ratio, angle of ion beam, and energy of ions. TEM images taken at various stages during the process seem to show that during the period following the initial stage the MgO formed cube shaped islands that did not touch until approaching compete coverage, and the orientation of each island is far from in-plane, perhaps 15° away. As they coalesce approaching complete coverage the angle is reduced greatly. This has been ascribed to grain-boundary energy minimization.

Further research at Stanford was not possible for a period (from 1998 till 2003). At Caltech Brewer and Atwater[13] in 2002 published their research using improved techniques in the art and science of RHEED together with TEM. This work differed from the Stanford mode by starting with a lower ion flux, somewhat less than the critical value. Thus the MgO deposited with little ion sputtering, with full coverage as an amorphous or micro-crystalline film. Suddenly at a thickness of about 20-30 Å the RHEED changed to a textured pattern.

THEORY: MOLECULAR DYNAMICS SIMULATION

At the same time period, Zepeda-Ruiz and Srolovitz (ZS) used atomistic simulations[14] to determine how an ion beam can control the crystallographic texture of a film by modifying the island nucleation process. They used the special case of MgO and 600 eV Ar^+ ion beams, and the substrate bed of amorphous SiN (Si_3N_4). As an illustration of the effect they compared two island sizes: one is 3x3x3 unit cell (about 1.3x1.3x1.3 nm for MgO), the other is 5x5x5 unit cell (2.1x2.1x2.1 nm for MgO). The result of a single Ar ion impinging along the channeling direction parallel to the (011) axis is dramatic: In the first case the island is completely amorphized, with some sputtering away of atoms, while the larger cell suffers very little damage and sputtering. Simulations for different angles of incident ion showed that the magic channeling direction is only effective for island sizes starting at about 2-3 nm, the differential with angle increasing as the size increased. A sobering finding is that the total sputtering yield minimum width is broad, about ±15 degrees (for the relatively low energies used in IBAD). This ultimately places a limitation on the expectations for biaxial alignment from this process alone.

They extended the simulations to include many ion impacts at variable angles, still for the special cases of 3x3x3, 5x5x5, and an infinite atom cube. To arrive at a description for all sizes of nuclei they apply an exponential fitting function to these three data points, arriving at the curves in Figure 1. The parameters are the size of the nuclei (film thickness as an approximation) and the ion / molecular flux ratio, ρ. These curves are for the case of Ar ions along the MgO (or any rocksalt-structure material) (011) channeling direction. They also show an example of the difference when the angle to the channeling direction is varied, with the result showing the dependence of the critical island size on island orientation with respect to the ion beam, i.e., where dL/dt =0. This figure is useful for discussing and describing the results of several of the well characterized published experiments, leading to some understanding and limitations of the basic understanding of the mechanism.

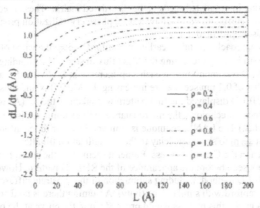

Figure 1. Variation of growth rate as function of island size L and ion / MgO ratio ρ. Taken from Zepeda-Ruiz, Srolovitz[14].

There are several issues to note about this simulation: This figure does not include the capillarity effect at small island size, which leads to the concept of a critical nuclei size even without the sputtering effect of the ions. This is evident in Figure 1 at the low ion flux and small island size. Another issue is: do the simulations mean actual sputtering of the atoms away from the growth region? The answer is yes, and this is defined where the atoms are moved 6 Å away from the surface. There is certainly damage (amortization) as well until grains crystallize when the conditions are right---in the region in Figure 1 above zero growth and until the size reaches about 3 nm.

One thing to note immediately is that the growth rate is zero for a ρ of about 0.55 (extrapolating where the curve passes through the origin). This is called the critical ρ_c, where there is no net deposition. This agrees well with the results of Caltech (Brewer-Atwater[13], Brewer thesis), and the Stanford results of Wang[11, 12]. However this does not agree with the results of Findikoglu-Matias[15, 16] (LANL), who find a value of greater than 1. We assume that there is a difference in the method of the determination of the fluxes. Caltech finds the best inplane alignment at a ρ of 0.43, thus as the shutter is opened there is a positive net island growth. If Figure 1 is followed the growth rate should follow the curve for the value of the ρ at the start and thus the rate of growth increases in time as the nuclei size increases, and thus the channeling differential increases. (We note here that we now measure the accumulation vs. time via a microbalance system---this will determine the time evolution of mass deposited, and check this assumption of the theory.)

Caltech used in-situ RHEED analysis and ex-situ TEM electron diffraction on the same SiN windows to monitor the development in real time as the thickness increases. They found that initially the deposit is amorphous/disordered fiber textured till a certain thickness of about 3 nm is reached, where suddenly the nuclei of biaxial textured MgO appear. The in-plane texture improves up to the thickness where the diffraction intensity is a maximum, and most of the film has transformed from amorphous to crystalline. They ascribe the texturing to the mechanism discussed above of ion beam minimal damage along the (011) direction and damage to the other orientations, resulting in orientated seeds in the disordered matrix. These seeds lead to a transformation of the surrounding region by a solid phase crystallization process, and is called in general due to "anisotropic ion damage".

A different approach was taken earlier at Stanford, where the value of ρ was experimentally determined by increasing the MgO flux in steps while holding the ion flux constant. The RHEED was monitored at each step-increase in MgO flux for a period of 20 sec (at an average MgO flux of 0.1 nm/sec) before increasing the MgO flux and thus decreasing ρ. The value where the RHEED displayed the first pattern was taken as the ρ_c. (As will be discussed later a new experimental technique, the microbalance, will examine the issue of is there actually deposition before this.) The LANL technique is similar. However the Stanford and the LANL experiments do not seem to show the delay in the crystallization till after a period of amorphous film growth, as seen in the Caltech process. Rather it seems that the amorphous/disordered fiber textured phase is not seen before the appearance of the RHEED pattern. However Caltech used a more sensitive RHEED method, including subtraction of the substrate diffuse pattern (this is now being repeated at Stanford with similar sensitivity). Assuming there is no disordered growth, using Figure 1 as a guide, this process would proceed along the curve at the ρ_c or slightly above (smaller ρ) till the size/thickness of 2-3 nm is reached and the channeling selectivity is in play. Note in this case the nuclei are separated from one-another; the starting point is a bare substrate. The higher ion beam flux used in the Stanford and LANL experiments have been argued to give rise to another mechanism, namely "anisotropic ion sputtering", rather than the Caltech "anisotropic ion damage". In both the Stanford and LANL experiments the smallest $\Delta\phi$ in-plane was found very close to ρ_c, supporting that proposal. Another supporting fact is that the value of the best in-plane $\Delta\phi$ is about 3° for the LANL, 6-7° for Stanford, and 10° for the Caltech. The reason for the larger value for the "anisotropic ion damage" process maybe that the seed nuclei have the range of alignment given by the ion channeling process, and as simulated by ZS, this has a range of ±15 degrees. As the seed grows using the neighboring amorphous material each seed-crystal is locked in place. However, in the Stanford work, where the starting condition is the bare substrate, TEM images taken at various stages in the growth seemed to show individual small crystallites separate from another and with this span of orientation, ± 15 degrees. However the final orientation at complete coverage has narrowed to 6-7°(and 3-4° for LANL). How this happens is still unknown, but a process of grain boundary energy minimization has been proposed. This may be possible if there is separation between the grains until the final completion of coverage is attained. This is aided if the grains are mobile, i.e., not strongly bonded to the bed-substrate.

We note that there is uncertainty as to when the improvement occurs. There is even speculation that this may occur during the heating up and/or during the homo-epi process. We propose to examine this carefully using the in-situ RHEED as well using ex-situ TEM, XRD (including synchrotron GIXRD), and AFM.

Sometime in the period 1998 the knowledge and techniques developed at Stanford were transferred to LANL by visits of staff member James (Randy) Groves, sharing unpublished data and Wang's thesis, and ongoing discussions. Groves and Paul Arendt [17] did extensive development of the MgO-IBAD as well as the "stack" of the other thin films that make up the package that makes for a successful Coated Conductor. This technology was then transferred to SuperPower, who have successively incorporated this technology with their MOCVD process of growing YBCO. SuperPower[18, 19] is now producing Coated Conductor at 360 m/hour. IBAD-MgO is now in production for Coated Conductor in Japan(Iijima[2] and Yamada[20]), and in Korea (S-I Yoo[21]).

NEW EFFORT AT STANFORD

Later a new effort was introduced at Stanford thanks to a grant from the Korean Electrotechnology Research Institute (KERI) in 2003. This made possible the design and construction of a new dedicated "Research on the Mechanisms of Ion Beam Assisted Deposition" facility. With the help of visiting students Wouter Baake of Twente University,The Netherlands, and Sung-Ho Hong of Seoul National University, Korea, this facility has been installed and is being used by a Materials Science student of Prof. Bruce Clemens, James (Randy) Groves.

In the planning for this new effort it was realized that two models for the MgO-texturing have evolved: The Caltech mode where there is no delay in the formation of the MgO film after the shutter is opened and the percentage of material sputtered away is small, and the Stanford mode, where most of the film is sputtered away, operating close to the critical ratio. Thus it seems important to monitor the mass accumulation in real time, while simultaneously monitoring the RHEED pattern at the same point of the film. A quartz crystal microbalance was designed and installed. Because the ion beam can deliver 100 mW/cm^2 of energy, good thermal heat transfer and bonding is important. For good ITaN growth and RHEED monitoring a smooth surface is required. The initial results are given in Groves' paper[5,6]. It is clear that the microbalance is giving us new and surprising results to consider.

Future Needs for In-situ Characterization of Morphology and Chemistry

Having these two in-situ characterization tools (RHEED and Microbalance) makes it clear that other characterization of morphology and chemistry are needed in order to make progress in our understanding and the control of the mechanisms of the IBAD-ITaN process. The in-situ tools are in addition to the ex-situ characterization being done:
- Ex-situ snap-shots during interrupted process:
  o TEM.
  o AFM.

- o SEM.
- o XPS.
- o XRD (including synchrotron GIXRD).

These ex-situ tools are important, but it is now clear that in-situ tools are very valuable. Some of the possible ones are:

- In-situ in real time:
  - o EELS-RHEED (diffracting species identification).
  - o Bi-axial texture analysis by RHEED (snap-shot if rotate substrate, or in real time if incorporate beam defection).
  - o Mass Spectrometer (Sputtered species).
  - o SEM (Installed in-situ).
  - o AFM (Installed in-situ but requires interruption for moving substrate).
  - o TOF-ISARS: Monitor composition of surface-when is coverage complete?
  - o LEEM.

EELES-RHEED appears very important----the chemical atomic species from the material doing the electron diffraction would be identified. In addition, the pass-energy feature would remove the RHEED electron scattering from the background and pass only the elastically diffracted electrons. This would greatly clarify the onset of different structures during the IBAD-ItaN process.

The Bi-axial texture analysis software is installed. At present it requires stopping the IBAD- ITaN process at stages and rotating the substrate by 10° while RHEED data is taken and analyzed. We do not know if this interruption is a perturbation to the growth process. An alternative that permits no interruption is to periodically deflect the angle of the beam by the same angle.

It is important that both of the latter two RHEED modifications are available commercially from Staib, the maker of our RHEED.

A mass spectrometer would have to be installed inside the chamber, without the normal flange. However this appears do-able. As we extend our research to elements in the bed-substrate and in the ITaN material the mass range of the spectrometer would have to cover these elements. Installing a SEM is feasible but challenging, but the benefit of observing the morphology in real time would be very important. The lower cost of a donated used system could make this possible. Installing an AFM inside the chamber (not totally in-situ) would be a challenge, including the mechanism for transferring the substrate to the AFM location.

TOF-ISARS would be easy to install and could be an important in monitoring the chemistry of the top surface. However it is very costly.

LEEM or in-situ TEM would be extremely interesting. The cost and physical requirements make this very difficult.

New Tools for Effecting Texture Development:

- Temperature, both during the initial stage and in the later stages. The earlier Stanford work found 300 °C not as good as at room temperature. This needs to be examined at all stages, for example in the later stage of coalescence as the grains align.
- Energy applied locally at the atomic level:
  - o Low energy ions, down to 15 volts is available.

o Photon assist during IBAD: The Rauschenbach group at the University of Augsburg in 1996-7 showed an effect of a Hg arc lamp, (with 500 mW/cm^2 at the substrate) on the TiN IBAD texture[22].
o There has been the suggestion of laser excitation of atomic levels to enhance the motion of atoms to lower energy positions.
o Electron beam heating, including that of the RHEED beam, has been shown to effect the crystallization of amorphous material[23].

Growth beyond the initial stage:

The effort is now focused on the initial stage of alignment. What about the development as complete coverage is approached?

- Caltech mode is assumed to have complete coverage from the beginning. When the thickness is ripe for the ion damage induced crystallization the seed is not free to move to minimize grain-boundary energy (locked in by complete coverage).
- Stanford mode is assumed to have isolated nuclei that survive and grow as separated islands. TEM shows them with large angles ($\pm$ 15). These are free to move on the bed till they connect and then grain-boundary energy minimization results in orientation to $\Delta \phi$ of 6-7° (3-4° for LANL).

The above models should have a dependence on the bond or interface energy between the MgO (in that case) and the "bed" material; say the SiN or a-Y2O3, etc., particularly in the Stanford case. Therefore the interface energy needs to be quantified, and tested by choosing different amorphous bed materials, such as SiN, Y2O3, Al2O3, SiOx, and a-metals (the conductivity could be a factor) and a-TiN, etc. Also the degree of ion etching on the bed could be a factor.

If we are right that the islands are free to move during the time before coalescence the prospect that the orientation might be influenced by an external force could be examined. One example might be an electric field acting on a certain IBAD material that is polar.

SUMMARY

- Increased understanding will come from in-situ characterization.
- Have only barely touched the possibilities:
    o Parameters include: ion/atom ratio, bed material and condition, temperature, ion energy, ion species.
    o New materials for the ITaN: oxides, nitrides, metals, etc.
    o New tools: photons, electrons, and multiple ion beams.
- Goals:
    o Understanding of the basic mechanisms of IBAD-ITaN.
    o Better alignment---what is the limit? Will application to semi-conductors, solar-cells be a possibility?
    o Faster and cheaper: Applications that depend on low cost processing (as is the case for Superconducting Coated Conductors, later for solar-cells)

ACKNOWLEDGEMENTS

The author thanks the many people who made the work possible: Prof. Ted Geballe who provided the environment for this research, Prof. Mac Beasley for his encouragement and facilities, Dr. Paul Grant who provided the original funding, saw the possibilities for IBAD-MgO, and his continued encouragement. This effort would not have been possible without the introduction to ion beams and their effect on thin film growth by Prof. Jim Harper, leading to the vision of ITaN. Khiem Do was instrumental in the design and assembly of the facility for the first effort at Stanford and in the original observation of the IBAD- ITaN on MgO, followed by Connie Wang who finished up the first phase, that now Randy Groves and Prof. Bruce Clemens are following up in the new effort. Important to the planning and understanding have been discussions with Prof and Dean David Srolovitz, Alp Findikoglu and Vladimir Matias, who also has been instrumental in obtaining funding through LANL and the Office of Electricity Delivery and Energy Reliability, DOE. The funding for the facility for the New Effort at Stanford is due to Profs Chan Park, Sung-Im Yoo, and William Jo, through KERI and the Center for Applied Superconductivity of the 21st Century Frontier R&D Program funded by the Ministry of Science and Technology, Republic of Korea.

REFERENCES

[1]   J. M. E. Harper, in Materials Research Society Fall Meeting 2008 (Boston, MA, 2008).

[2]   Y. Iijima, in Materials Research Society Fall Meeting 2008 (Boston, MA, 2008).

[3]   R. H. Hammond, in Materials Research Society Fall Meeting 2008 (Boston, MA, 2008).

[4]   R. M. Bradley, J. M. E. Harper and D. A. Smith, Journal of Applied Physics 60 (1986) 4160.

[5]   J. R. Groves, R. F. DePaula, L. Stan, R. H. Hammond and B. C. Clemens, IEEE Transactions on Applied Superconductivity (2009) in press.

[6]   J. R. Groves, in Materials Research Society Fall Meeting 2008 (Boston MA, 2008).

[7]   Y. Iijima, N. Tanabe, O. Kohno and Y. Ikeno, Applied Physics Letters 60 (1992) 769.

[8]   X. D. Wu, S. R. Foltyn, P. Arendt, J. Townsend, C. Adams, I. H. Campbell, P. Tiwari, Y. Coulter and D. E. Peterson, Applied Physics Letters 65 (1994) 1961.

[9]   P. N. Arendt, S. R. Foltyn, J. R. Groves, R. F. DePaula, P. C. Dowden, J. M. Roper and J. Y. Coulter, Applied Superconductivity 4 (1996) 429.

[10]   R. P. Reade, P. Berdahl, R. E. Russo and S. M. Garrison, Applied Physics Letters 61 (1992) 2231.

[11]   C. P. Wang, Doctor of Philosophy, Materials Science and Engineering, (Stanford University, 1999).

[12]   C. P. Wang, K. B. Do, M. R. Beasley, T. H. Geballe and R. H. Hammond, Applied Physics Letters 71 (1997) 2955.

[13]   R. T. Brewer and H. A. Atwater, Applied Physics Letters 80 (2002) 3388.

[14]   L. A. Zepeda-Ruiz and D. J. Srolovitz, Journal of Applied Physics 91 (2002) 10169.

[15]   A. T. Findikoglu, J. Mater. Res. 19 (2004) 501.

[16]   V. Matias, J. Hanisch, E. J. Rowley and K. Guth, Journal of Materials Research 24 (2009) 125.

[17]   J. R. Groves, P. N. Arendt, S. R. Foltyn, Q. X. Jia, T. G. Holesinger, H. Kung, E. J. Peterson, R. F. DePaula, P. C. Dowden, L. Stan and L. A. Emmert, Journal of Materials Research 16 (2001) 2175.

[18]   V. Selvamanickam, Y. Chen, X. Xiong, Y. Xie, X. Zhang, Y. Qiao, J. Reeves, A. Rar, R. Schmidt and K. Lenseth, Physica C: Superconductivity 463-465 (2007) 482.
[19]   X. Xiong, in Materials Research Society Fall Meeting (Boston MA, 2008).
[20]   Y. Yamada, in Materials Research Society Fall Meeting (Boston, MA, 2008).
[21]   S.-I. Yoo, in Materials Research Society Fall Meeting (Boston, MA, 2008).
[22]   M. Zeitler, J. W. Gerlach, T. Kraus and B. Rauschenbach, Applied Physics Letters 70 (1997) 1254.
[23]   A. Sasaki, H. Isa, J. Liu, S. Akiba, T. Hanada and M. Yoshimoto, Japanese Journal of Applied Physics 41 (2002) 6534.

[18] V. Sandoghar, K. Y. Chen, Y. Kudyk, Y. Guo, Y. Zhang, W. D. Oliver, and A. Kapitulnik, Applied Physics C. Superconductivity 46–50 (2000/12).

[19] A. Aharony, Research Society PhD Meeting, Boston MA, 2000.

[20] W. Zhang, Materials Research Society PhD Meeting, 1999.

[21] R. J. Cava, Materials Research Society Fall Meeting, Boston, 2000.

[22] M. Zaffe, J. W. Lynn, J. Zhou, and L. Raymond, etc. Appl. Phys. Magazine, 11(2000) 234.

[23] A. Fasselkof, L. Tan, N. Mayer, P. Banks, and D. Whitingnet Superconductor Physics 41 (2000) 678.

IBAD Texturing

Mater. Res. Soc. Symp. Proc. Vol. 1150 © 2009 Materials Research Society   1150-RR02-02

Investigation of Early Nucleation Events in Magnesium Oxide During Ion Beam Assisted Deposition

James R. Groves[1], Robert H. Hammond[2], Raymond F. Depaula[3], and Bruce M. Clemens[1]

[1]Department of Materials Science and Engineering, Stanford University, Stanford CA 94305
[2]Geballe Laboratory for Advanced Materials, Stanford University, Stanford CA 94305
[3]Superconductivity Technology Center, Los Alamos National Laboratory, Los Alamos, NM 87545

ABSTRACT

Ion beam assisted deposition (IBAD) is used to biaxially texture magnesium oxide (MgO), which is useful as a template for the heteroepitaxial growth of various thin film devices and most notably as a template layer for high temperature superconductors. Improvements in the quality of IBAD MgO films have been largely empirical and there is uncertainty as to the exact mechanism by which this biaxial texture is developed. Using a specially built quartz crystal microbalance (QCM) as both a substrate and monitor in conjunction with reflected high-energy electron diffraction (RHEED) acting on the same surface, we have probed the initial stages of IBAD MgO growth in-situ. We have correlated corresponding RHEED images with real-time mass accumulation QCM data during the film growth. During IBAD growth, the mass accumulation exhibits a sharp change in slope corresponding to a sudden decrease in growth rate. Corresponding RHEED images show an abrupt onset of crystallographic texture at this point. A simple model incorporating differential etch rates of the MgO film and silicon nitride substrate can be used to fit the data but is inconsistent with the behavior during ion etching with no growth. It is, therefore, postulated that a more complex mechanism is responsible for the observed behavior.

INTRODUCTION

The ion beam assisted deposition (IBAD) technique has been used to develop biaxial texture in a number of metals, metal oxides and metal nitrides[1-4]. The process is an enabling method for the development of second generation high-temperature superconducting coated conductors [2]. Initially, IBAD yttria-stablized zirconia (YSZ) allowed researchers to deposit bi-axially aligned template films for the subsequent deposition of heteroepitaxial thin films of $YBa_2Cu_3O_{7-\delta}$ (YBCO) on flexible metal substrates [5-7]. However, these IBAD YSZ films were prohibitive in that they required excessive processing times that precluded them from adaptation to continuous processing for industrial scale-up [8]. In response to this requirement, researchers at Stanford University found that IBAD could be applied to magnesium oxide (MgO) with greater efficiency [9]. MgO required only 10 nm to develop comparable in-plane texturing whereas IBAD YSZ needed at least 0.5 μm [10].

Researchers have continued to improve the IBAD MgO process and adapt it to continuous processing. Most notably, researchers at Los Alamos National Laboratory (LANL) have succeeded in improving the in-plane alignment of biaxially aligned polycrystalline IBAD

MgO films with phi-scan full-width at half-maximum (FWHM) values of 5° having been routinely achieved on continuously processed metal tape [11]. These improvements have consisted of developing new IBAD MgO nucleation layers and optimizing ion gun processing parameters. However, even the noteworthy long-length performance of SuperPower, Inc. has been largely based on an empirically determined deposition technique [12].

Several theories have been proposed to explain the mechanism of IBAD MgO texture development. Three primary models have been widely accepted by the IBAD community. Wang et al. propose that oriented grains survive due to differential ion beam sputtering and, upon coalescence; these grains undergo a minimization of grain boundary energy that results in grain-to-grain alignment through relative grain rotations [13]. The second model, proposed by Usov and co-workers, depends upon damage anisotropy between crystallographic planes in the grains [14]. Orientations with higher tolerances for damage accumulation survive and grow at the expense of other grains resulting in a preferred orientation. Finally, Brewer and Atwater theorize that oriented grains grow through induced solid phase crystallization by ion beam bombardment of an initially amorphous MgO layer [15].

In a recent paper, we presented data using a novel in-situ quartz crystal microbalance (QCM) as a substrate during IBAD MgO growth [16]. Experiments using this in-situ QCM with temporally correlated reflected high-energy electron diffraction (RHEED) images showed an intriguing phenomenon in the thickness versus time data in a series of experiments using different ion-to-molecule ratios. In each case, the mass accumulation exhibits a sharp change in slope indicating a sudden decrease in growth rate at a critical thickness of ~2 nm. Correlation with RHEED images taken during the deposition showed that MgO developed in-plane crystallographic texture at this same critical thickness.

In this paper, we present data collected on the nucleation events of IBAD MgO. Here, we have further improved our in-situ method to measure the mass accumulation of MgO during IBAD growth with concurrent, temporally correlated monitoring by RHEED. The development of IBAD MgO texture and its correlation to RHEED patterns are discussed in terms of an improved model that describes the deposition data as a function of mass density rather than the observed thickness.

EXPERIMENT

Optically polished 50 nm gold-coated 14 mm diameter 5 MHz quartz crystals (< 3 nm rms) were purchased from Q-sense, Inc. (Glen Burnie, Maryland). These as-received crystals were placed in a custom made fixture. The fixture consisted of a water-cooled copper block that provided sufficient thermal contact with the oscillating crystal using a thin indium seal. The quartz crystal was used as both a QCM and substrate for subsequent depositions of a nucleation layer and the IBAD MgO film. A second integrated QCM crystal was used as the monitor and control for the deposition rate of the MgO vapor flux.

All depositions were conducted in a high vacuum chamber with a typical base pressure of 7.0×10^{-6} Pa (5.0×10^{-8} torr) at room temperature. A four-pocket 7 cc Temescal SuperSource provided the deposit vapor flux. A two-grid collimated Kaufman ion source at an incidence angle of 45° relative to the substrate normal provided an Ar ion flux to the substrate. The ion fluence was monitored with a separate Faraday cup. The Faraday cup was biased at -100 V to eliminate contributions from electrons to the ion current reading.

Several different thin films were deposited directly onto the in-situ Au-coated quartz crystal and used as IBAD MgO nucleation layers. Typically, a 10 to 20 nm thick amorphous layer of Si_3N_4 was used. No background N_2 gas was introduced into the chamber during the Si_3N_4 deposition. The crystalline nature of the deposit was confirmed by observation of the RHEED pattern during deposition.

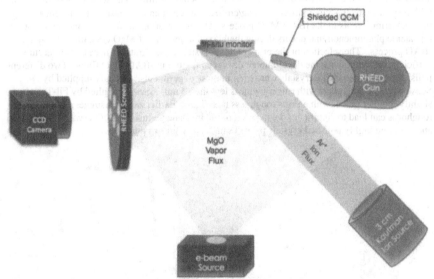

Figure 1. A schematic of IBAD chamber used for in-situ QCM experiments. Vapor flux is provided by an offset e- beam hearth. The Kaufman-type ion source is angled at 45° relative to the substrate normal and a RHEED beam impinges perpendicular to the ion gun axis. The in-situ monitor has an additional shielded QCM monitor to measure the MgO vapor flux.

The IBAD MgO layer was deposited with concurrent 750 eV Ar ion and MgO fluxes. An electron beam evaporator provided the MgO vapor flux at 0.05 nm/s to 0.10 nm/s. The flow rate of Ar gas into the system was kept constant at 10 sccm, which corresponded to a chamber pressure of $\sim 5.0 \times 10^{-3}$ Pa. The ion to atom ratio was varied to determine its effect on nucleation of the MgO films. IBAD film growth was monitored in-situ using RHEED. Frames were taken at 1-second intervals during deposition. The RHEED beam is aligned along an axis 90° relative to the ion beam. All patterns were taken at a beam energy of 25 keV.

A computer interface allowed the simultaneous collection of data from both the in-situ and shielded QCMs as well as the images from the camera monitoring the RHEED screen. Direct correlation between the mass accumulation data (and thickness) of the in-situ QCM and the RHEED images were made as a function of time. A schematic of the deposition chamber is shown in Figure 1 with the various instrumentation and apparatus labeled.

DISCUSSION

In-situ QCM data

The use of the in-situ QCM monitor as both a substrate and mass accumulation sensor allows us to record the growth of IBAD MgO as a function of time. The sensitivity of the QCM with respect to mass accumulation (~ 20 ng/cm^2·Hz) makes it an ideal instrument for measuring the nucleation events in the IBAD MgO process. Our observations have found that there is a repeatable phenomenon that occurs during the biaxial alignment of MgO crystallites using the IBAD process. The selection of the proper QCM crystals as substrates is necessary to ensure sufficient smoothness for the development of biaxial texture in IBAD MgO films. Two different polished, Au-coated QCM crystals were tried in these experiments. Crystals supplied by Q-Sense were the smoothest with rms roughness less than 3 nm. Crystal supplied by Fil-Tech had a slightly rougher polish with an rms roughness near 7 nm. Earlier research suggested that roughness can lead to significant degradation of the in-plane texture [17]. Both crystals were tested and the more highly polished crystals proved satisfactory for our purposes.

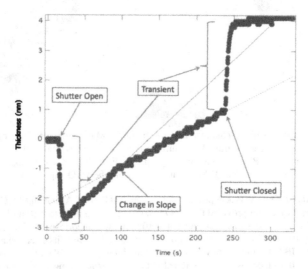

Figure 2. A graph showing the thickness versus time data for a typical IBAD MgO film growth on a polished in-situ QCM crystal.

Figure 2 shows the experimentally observed mass accumulation behavior for IBAD growth on a QCM. The in-situ QCM exhibits two different features. Firstly, there are transients at the beginning and end of growth associated with the onset of exposure of the QCM to the ion and growth flux. These transients are most likely the result of heating of the crystal during ion bombardment. Secondly, there is a change in slope, corresponding to a change in growth rate during the film growth at a critical thickness.

Previously, we also found that data collected for a series of experiments at different ion-to-molecule ratios exhibited the similar two-slope behavior observed in our initial experiments. In each case, as the ion-to-molecule ratio is increased, the observed slopes decrease in magnitude. This is as expected since the ion beam current density is increasing leading to a greater etch rate. However, the change in slope occurs at a similar point (~ 2 nm thickness) in each case, regardless of the ratio value. This suggests that there is a critical thickness for the nucleation of the biaxially oriented MgO grains as postulated by Zepeda-Ruiz and Srolovitz[18]. Such behavior was also suggested by Im and Atwater[19] for the ion irradiation enhanced nucleation of crystalline Si in amorphous Si thin films and also proposed by Brewer and Atwater as the explanation for the development of texture in IBAD MgO[15].

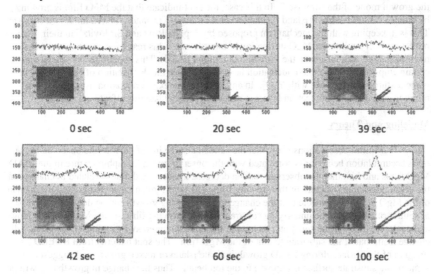

Figure 3. Composite images for an IBAD MgO deposition on the in-situ QCM as a function of deposition time. The top graph in each composite set is a line profile taken through the (02) spot at the bottom center of the RHEED image in the lower left corner of each set. The lower right graph in each set shows the monitor QCM (light trace) and the in-situ QCM (dark trace) thickness values.

QCM data and temporally correlated RHEED images were taken for a series of IBAD MgO thin film experiments. Figure 3 shows a set of snapshots taken for one such experiment. In this case a 10 nm thick amorphous Si_3N_4 was used as a nucleation layer, and the IBAD MgO film was deposited using a vapor rate of 0.07 nm/s and ion beam flux of 70 μA/cm^2. The series of images shown in Fig. 3 were taken at six different times in the deposition. Each set contains a line scan through the (02) diffraction spot, a RHEED image and the accompanying QCM data for both the monitor and in-situ crystals. Initially, only a broad background is observed, which is indicative of the amorphous Si_3N_4 substrate layer. After 20 seconds of IBAD MgO growth, the RHEED pattern shows the formation of rings indicating the development of polycrystalline MgO grains. These images illustrate the abruptness of the onset of the crystallographic texture and its correlation with sudden decrease in growth rate. The spot RHEED pattern associated with in-plane crystallographic orientation appears between 39 and 45 seconds in Fig. 3. At this same time (~ 40-45 seconds) the kink appears in the mass accumulation data, indicating that the net growth rate decreases sharply as the crystallographic texture develops.

The data indicates that the development of biaxial texture in IBAD MgO occurs rapidly as a critical thickness in the film is reached. Initially, randomly oriented MgO crystallites are deposited on the surface. Once a sufficient thickness or island size is present, the MgO crystallites begin to orient. Further evidence of this process is given by the RHEED image. The spot pattern is not only indicative of the biaxial texture in the developing film but also describes the growth mode of the film itself. In this case, the spot indicate that the MgO film is growing by either a Volmer-Weber (island formation) or a Stranski-Krastanov (mixed mode) mechanism. This is in keeping with the mechanism proposed by Zepeda-Ruiz and Srolovitz[18] in their modeling of crystallites of MgO during ion bombardment. This result further suggests that the nucleation layer can influence the growth mode of the IBAD. This is perhaps the reason that certain empirically determined nucleation layers can reduce the degradation of IBAD MgO in-plane texture, as is the case with Y_2O_3. In any event, more experimentation using the in-situ monitor with different nucleation layers is necessary to elucidate this effect.

Modeling and Theory

Our initial observations show that the sudden change in net growth rate observed in the mass accumulation behavior is correlated with the onset of crystallographic texture in the IBAD MgO films and hint that the observed sudden decrease in net growth rate is related to changes in the crystallographic structure of the growing film. However, given the island growth mechanism of the MgO, it is also possible that the change in slope during growth can be due to a difference in sputter rate between the underlying substrate and the growing film. As the MgO islands grow and coalesce, to cover the substrate, the rate of material sputtered away by the Ar ion beam will change from that of the substrate to that of the MgO film. The spot patterns from RHEED imaging of the surface during IBAD growth indicate island or mixed growth and suggest that some of the substrate surface is exposed to the ion beam. Thus the change in growth rate will be due to coverage of the substrate by the growing film rather than changes in the crystallographic structure of the film.

To investigate this possibility we developed a model incorporating different sputter rates for the film and substrate during growth, and a critical film thickness where the islands agglomerate and the sputter rate changes from that of the substrate to that of the film. During the initial growth, the net growth rate is a balance of the film deposition flux and the ion etch rates of the film. The growth rate is then given by:

$$\frac{dh}{dt} = V_F \left(J_F - J_I S_F \frac{h}{h_c} \right)$$

where h_c is the thickness associated with complete coverage of the substrate by the film, J_F is the deposition flux of film material, J_I is the ion beam flux, S_F is the sputter yield of the film, and V_F

is the molecular volume of the film. During this initial growth, the quantity h/h_c represents the fraction of the substrate covered by the film. For $h > h_c$ the substrate is completely covered by the film and the growth rate is given by:

$$\frac{dh}{dt} = V_F(J_F - J_I S_F)$$

Integrating, we find the film thickness is given by:

$$h = \begin{cases} \frac{h_c J_F}{J_I S_F}\left[1 - \exp\left(\frac{-J_I S_F V_F t}{h_c}\right)\right] & h < h_c \\ h_c + V_F(J_F - J_I S_F)(t - t_c) & h > h_c \end{cases}$$

where t_c is the time at which full film coverage is achieved, and is given by:

$$t_c = \frac{-h_c}{J_I S_F V_F} \ln\left(1 - \frac{J_I S_F}{J_F}\right)$$

The QCM measurement during IBAD film growth reflects the decrease in mass due to sputtering away of the substrate and film as well as the increase of mass due to the depositing film material. Taking this into account we find the rate of increase in mass during the initial deposition ($h < h_c$) is given by:

$$\frac{1}{A}\frac{dM}{dt} = J_F M_F - J_I\left[S_F M_F \frac{h}{h_c} + S_S M_S\left(1 - \frac{h}{h_c}\right)\right]$$

where A is the area of the QCM, M_F and M_S are the molar masses for the film and substrate material, S_S is the sputter rate of the substrate material, and h is the film thickness as calculated above. Integrating we find (for $h < h_c$):

$$\frac{M}{A} = (J_F M_F - J_I S_S M_S)t - \left(\frac{S_F M_F - S_S M_S}{S_F}\right)\left(J_F t - \frac{h(t)}{V_F}\right)$$

For $h > h_c$ the mass accumulation is just a balance of the deposition and sputter rate of the film, and is linear in time:

$$\frac{M}{A} = M_F[h_c + (J_F - J_I S_F)(t - t_c)]$$

Figure 4 shows the results of fitting the QCM data to the above model for the case of deposition of an MgO film on amorphous Si_3N_4, where we have used the data from the monitor QCM (which was exposed to the same growth flux but no ion beam flux) to determine the film deposition rate. The fit to the observed behavior is excellent and the model reproduces the change in slope. The mass uptake behavior is consistent with a MgO film that reaches complete coverage at ~2.7 nm and which has a sputter rate about 3 times higher than that of the Si_3N_4 substrate layer. Thus as the substrate is covered by the growing MgO film, the net growth rate is reduced due to the higher etch rate of the MgO compared to the Si_3N_4 substrate layer.

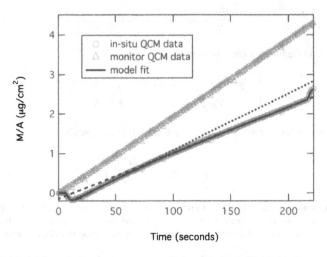

Figure 4. Model fit to the in-situ mass accumulation data for an IBAD MgO growth.

In order to confirm this behavior we conducted etch experiments to independently measure the sputter etch rate for MgO and Si_3N_4 under conditions similar to those during growth, A bilayer of 20 nm of polycrystalline MgO was deposited on top of 20 nm of amorphous Si_3N_4 deposited on the in-situ QCM, and the resulting bilayer was etched using a constant ion beam flux. As shown in Fig. 5 there is a clear change in slope as the MgO film is etched away and the ion beam begins etching the Si_3N_4 substrate layer. However, the MgO etch rate (found from the initial slope in Fig. 5) is *less* than that for the Si_3N_4 substrate layer (0.012 $\mu g/cm^2 s$ vs 0.016 $\mu g/cm^2 s$, respectively). This is inconsistent with the mass uptake during growth, where the etch rate of the MgO film is *larger* than that of the Si_3N_4 substrate layer. Thus, the observed kink in the mass uptake behavior during growth cannot be due solely to the differential sputter etch rates and the coverage of the substrate by a film. This suggests that the observation that the sudden change in mass uptake is simultaneous with the abrupt appearance in film crystallographic texture is not just a coincidence, and that the abrupt decrease in mass uptake is related to the change in the crystallographic structure of the film, not just film coverage.

The origin of the connection between structure and net growth rate could be related to the surface structure of the growing film. Before texture is developed, the film is highly disordered and the surface likely has a high concentration of kink sites favorable for attachment of arriving MgO. Thus MgO adatoms are rapidly incorporated into the growing film and are more resistant to being sputtered away by the ion flux. After the crystallographic texture develops, the film surface is likely more perfect with fewer kink sites (this is specially true if the development of in-plane crystallographic texture is associated with agglomeration), and thus adatoms travel a greater distance before encountering a kink site and being incorporated into the film. During this time they are more easily sputtered away and will thus have a lower net sticking coefficient than the case for the more disordered film. We note that the model we have constructed above is applicable to this situation as well, with the substitution of $J_F = 0$. Future experiments are planned to investigate this, including using a mass spectrometer to measure the presence of sputtered species.

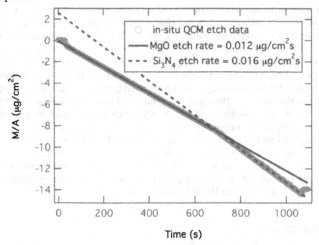

Figure 5. In-situ QCM data shown as a function of time during a bilayer etch of polycrystalline MgO (20 nm thick) and amorphous Si_3N_4 (20 nm thick). Linear fitting of the two slopes shows that the polycrystalline MgO etches more slowly than the amorphous Si_3N_4.

Although the above mechanism for connecting the film structure with the net growth rate is highly speculative, we can make some concrete conclusions about this behavior and the mechanism of ion beam texturing. Most models for development of crystallographic texture during IBAD growth rely on a differential in ion damage between favorably and unfavorably oriented islands or regions of the film. Favorably oriented regions or islands survive the ion bombardment and thus dominate, while randomly oriented regions or islands are destroyed and re-sputtered by the ion bombardment. This would predict that the net growth rate of the film would *increase* as the film develops a favorable texture. The observation of the opposite behavior challenges the basic assumptions of this class of model and calls for a re-examination of the origins of texture development during IBAD growth.

It could be argued that the incorporation of argon ions in the growing lattice affects the microstructure of the film. Many experiments have looked at the effect of inert ion entrapment in both thin films and bulk crystals[20]. The majority of these experiments, however, suggest that although the inert gas penetrates into the surface of the film and that this is a function of the ion energy, the trapping probability is very low (<0.1%) for ions with energies below 1 keV[21]. Furthermore, we expect that, due to the islanded growth of IBAD MgO films, the diffusion path for desorption of Ar atoms is small and most likely reduces the entrapment probability even more. As the film thickness increases, the entrapment probability will increase but remains small and, therefore, we expect that the influence of entrained Ar is also small.

SUMMARY

We have used a specially built quartz crystal microbalance (QCM) as both a substrate and mass accumulation measurement device during IBAD MgO film growth. Temporally correlated RHEED images were also obtained. We observe a sudden decrease in mass accumulation rate at the same time as an abrupt appearance of crystallographic texture. The decrease in net growth rate and the development of biaxially crystallographic texture occur at a film thickness of ~2-3 nm. A simple physical model that ascribes the observed behavior to the effect of coverage of the Si_3N_4 substrate by the MgO film can duplicate the observed behavior if the MgO sputter yield is about 3 times that of the substrate. However, this sputter yield behavior is inconsistent with separate ion etch experiments, which reveal that the MgO etches *slower* than the amorphous Si_3N_4 layer. We suggest that the origin of the decrease in mass uptake is related to the simultaneous change in film crystallographic structure and we further postulate that the crystallographic texture results in an increase in film surface perfection that leads to a greater difficulty in incorporating adatoms into the growing film. In any case, the observation of a greater sputter yield associated with the development of crystallographic texture challenges the basic assumptions of most models for IBAD induced texture.

ACKNOWLEDGMENTS

The authors wish to thank the generous funding from both the Office of Electricity Delivery and Energy Reliability, U.S. Department of Energy and from the Center for Applied Superconductivity Technology of the 21st Century Frontier R&D Program funded by the Ministry of Science and Technology, Republic of Korea through a grant to Stanford University.

REFERENCES

1. L. S. Yu, J. M. E. Harper, J. J. Cuomo and D. A. Smith, Appl. Phys. Lett. **47** (9), 932 (1985).
2. P. N. Arendt and S. R. Foltyn, MRS Bulletin **29** (8), 543 (2004).
3. W. Ensinger and M. Kiuchi, Surf. Coat. Tech. **84** (1/3), 425 (1996).
4. J. W. Gerlach, U. Preckwinkel, H. Wengenmair, T. Kraus and B. Rauschenbach, Appl. Phys. Lett. **68** (17), 2360-2362 (1996).
5. Y. Iijima, N. Tanabe, O. Kohno and Y. Ikeno, Appl. Phys. Lett. **60** (6), 769-771 (1992).
6. R. P. Reade, P. Berdahl, R. E. Russo and S. M. Garrison, Appl. Phys. Lett. **61** (18), 2231-2233 (1992).
7. X. D. Wu, S. R. Foltyn, P. Arendt, J. Townsend, C. Adams, I. H. Campbell, P. Tiwari, Y. Coulter and D. E. Peterson, Appl. Phys. Lett. **65** (15), 1961-1963 (1994).
8. D. K. Finnemore, K. E. Gray, M. P. Maley, D. O. Welch, D. K. Christen and D. M. Kroeger, Physica C **320** (1-2), 1-8 (1999).

9.   C. P. Wang, K. B. Do, M. R. Beasley, T. H. Geballe and R. H. Hammond, Appl. Phys. Lett. **71** (20), 2955-2957 (1997).

10.   P. N. Arendt, S. R. Foltyn, J. R. Groves, R. F. DePaula, P. C. Dowden, J. M. Roper and J. Y. Coulter, Appl. Supercond. **4** (10-11), 429-434 (1996).

11.   P. N. Arendt, S. R. Foltyn, L. Civale, R. F. DePaula, P. C. Dowden, J. R. Groves, T. G. Holesinger, Q. X. Jia, S. Kreiskott, L. Stan, I. Usov, H. Wang and J. Y. Coulter, Physica C **412/414P2**, 795-800 (2004).

12.   V. Selvamanickam, Y. Chen, X. Xiong, Y. Xie, X. Zhang, Y. Qiao, J. Reeves, A. Rar, R. Schmidt and K. Lenseth, Physica C **463-465**, 482-487 (2007).

13.   C. P. Wang, Doctor of Philosophy, Stanford University, 1999.

14.   I. O. Usov, P. N. Arendt, J. R. Groves, L. Stan and R. F. DePaula, NIM B **243** (1), 87 (2006).

15.   R. T. Brewer and H. A. Atwater, Appl. Phys. Lett. **80** (18), 3388-3390 (2002).

16.   J. R. Groves, R. F. DePaula, L. Stan, R. H. Hammond and B. C. Clemens, IEEE Trans. Appl. Supercond., in press (2009).

17.   J. R. Groves, P. N. Arendt, S. R. Foltyn, Q. X. Jia, T. G. Holesinger, H. Kung, E. J. Peterson, R. F. DePaula, P. C. Dowden, L. Stan and L. A. Emmert, J. Mater. Res. **16** (8), 2175-2178 (2001).

18.   L. A. Zepeda-Ruiz and D. J. Srolovitz, JAP **91** (12), 10169-10180 (2002).

19.   J. S. Im and H. A. Atwater, Appl. Phys. Lett. **57** (17), 1766-1768 (1990).

20.   H. F. Winters, H. Coufal, C. T. Rettner and D. S. Bethune, Phys. Rev. B **41** (10), 6240 (1990).

21.   E. V. Kornelsen and M. K. Sinha, JAP **39** (10), 4546-4555 (1968).

Mater. Res. Soc. Symp. Proc. Vol. 1150 © 2009 Materials Research Society

Generating MgO single- and bi-crystal templates on hard and soft substrates using ion beam assisted texturing process

Judy Z. Wu, Rongtao Lu and Ronald N. Vallejo
Department of Physics and Astronomy, University of Kansas, Lawrence, KS 66045, USA

ABSTRACT

Ion beam assisted texturing process (IBAD) has provided a promising approach for development of epitaxial thin film devices on various low-cost nontextured substrates. The interface between the IBAD template and the substrate surface has been found to play the key role in texture quality of the template. In this work, we have investigated texture evolution of IBAD MgO templates on several technological important substrates including cereflex (hard) and polyimide (soft). The former may be an alternative of low AC loss substrates for high-T_c superconductor tapes but has extremely rough surface incompatible to IBAD MgO process. By developing a surface smoothening process, high-quality bi-axially textured MgO templates have been achieved on Ceraflex. On the amorphous polyimide films, preferential sputtering of the ion beam prevents texturing of IBAD MgO. With thin buffer layer on top of the polyimide surface, the preferential sputtering of polyimide surface can be minimized and highly textured MgO template with in-plane full-width-at-half-maximum of 9-10° and out-of-plane full-width-at-half-maximum~3.0° have been obtained. This method may provide a practical route for fabricating suspended epitaxial devices on polymer sacrificial layers as demonstrated. In addition, a two-step process was developed to generate textured MgO bicrystal templates on glass and Si substrates. After the first-step IBAD process, a mask was generated on the MgO template and a selected area was removed before a second MgO template was laid at a selected in-plane angle. The flexibility of this technique makes it possible to fabricate various two dimensional novel bicrystal devices of microscopic scales.

INTRODUCTION

Epitaxial thin films are the foundation for various solid-state devices including transistors, diodes, lasers, sensors, etc. This prompts an extensive research during past many in fabrication and application of thin films of a large spectrum of materials including semiconductors, superconductors, ferroelectric, dielectric and magnetic materials. It should be realized that many physical properties can be obtained only if single crystalline thin films in which atoms align into a perfect crystalline lattice. This makes thin film epitaxy both technologically and fundamentally important and critical for solid-state devices and integrated circuits. Epitaxy of thin films is generally obtained on single crystalline substrates via direct epitaxy or hetero-epitaxy. The crystalline quality of the film is sensitively dictated by factors such as surface morphology of the substrate, lattice mismatch and chemical compatibility between the film and the substrate. This in many cases results in limited choices of substrates for epitaxy of certain materials. In addition, single crystalline substrates are often very expensive and may not be available in large size. In contrast, many low cost substrates available in large size, such as metals, ceramics, glass, polymers, either have no texture or have polycrystalline texture. In many practical applications, in fact, epitaxial films are required on these non-textured substrates including amorphous and polycrystalline ones. This justifies the need for textured templates that enable epitaxy of thin or thick films of technologically interesting materials.

Ion beam assisted deposition (IBAD) combines a traditional deposition technique and the simultaneous irradiation of the growing film with ions at a certain angle from the substrate normal. The simultaneous use of ion beam during deposition has been found to induce texture formation during thin film growth. Early studies have shown that ion beam assisted growth of Nb films develops in-plane crystallographic texture parallel to the ion beam direction while Nb films grown without ion assistance develop no in-plane texture[1]. Recently, ion beam assisted deposition has been adopted to generate bi-axially textured templates. Among few others, popular template materials include yttria-stabilized zirconia (YSZ)[2-5] and magnesium oxide (MgO)[6-18] which have enabled hetero-epitaxy of high temperature superconductors on non-textured metallic substrates. The texturing of YSZ, however, occurs rather slowly, in that it requires > 700 nm thickness of film to achieve good in-plane alignment. In contrast, previous reports[6,8-10, 13] on bi-axially textured MgO templates indicate that texturing in films as thin as 10 nm is formed, thereby reducing the processing time by a factor of 100 as compared to that of YSZ. The rapid bi-axial texture formation in MgO makes it technologically interesting for various device applications.

The mechanism of this texturing process remains under debate. The major difference in the models proposed is in where (or at what thickness) the texture initiates in MgO layer under a specific IBAD processing condition. At a higher ion-to-atom ratio most published work employs, it is generally believed that the texture is established at the nucleation stage before the MgO islands coalesce into a continuous film[8]. This makes the substrate original surface condition and film-substrate interaction during the initial stages of the film growth critical components in the texture evolution of MgO films[11]. At a lower ion-to-atom ratio, a critical MgO thickness on the order of few nm is argued necessary before texture appears[9]. On the other hand, the texture quality remains a topic despite constant progress made by optimizing various processing parameters including collimation of the ion beam, ion-to-atom ratio, buffer layers, etc. The in-plane spread of grain alignment of several degrees, usually measured in x-ray diffraction (XRD) full-width-at-half-maximum, remains to be improved as required for many applications of semiconductors.

Besides interesting physics in nucleation mechanism of the texture in MgO, exploration of the IBAD texturing approach for different applications is also important. In recent years, we have worked on several technologically interesting substrates including hard substrates like glass[15], SiO_2/Si[15], ceraflex[16-18] and soft substrates such as polyimide and photoresist. Glass and SiO_2/Si substrates are among the most popular substrates for various semiconductor and photonic devices due to their low price and availability in large dimension. Ceraflex is a promising substrate candidate for low ac loss coated conductors of high T_c superconductors with high resistivity on the order of few hundreds $\Omega\cdot cm$[5] and can be made flexible. For 100 μm thick Ceraflex tapes, the radius of curvature is approximately 8 mm. Polymers like polyimide and photoresist may be used directly as substrates and also are widely used in microelectronic device processing. To be able to develop IBAD-MgO textured templates on these soft substrates may provide promising potential in applications of sensors, detectors, and displays. One particular question we have been trying to ask is what serves as the adequate condition on a substrate surface for high quality IBAD-MgO textured template to develop. In this work, we shed some light on this based on our recent work on both hard and soft substrates. In addition to single crystalline IBAD MgO template, we have explored bi-crystal templates on SiO_2/Si[19] and possible improvements and remaining issues in this direction will be discussed.

EXPERIMENT

An electron beam evaporation system equipped with a 3 cm Kauffman ion source was used for IBAD-MgO texturing process. The experimental details were reported earlier[15-19]. Briefly, an Ar^+ beam of 750 eV and 10 mA as measured by a Faraday cup near the substrate was directed at 45° with respect to the normal of sample stage, which is a (110) channeling direction for (100) oriented MgO. In some cases, Y_2O_3 amorphous buffer layer was deposited on substrates before ion-beam assisted e-beam deposition of 10-15 nm MgO textured templates at room temperature. The ion-to-atom ratio was chosen to be ~0.9 for the IBAD texturing process. A homo-epitaxial MgO layer was then grown at 400°C for the convenience of texture quality examination. The *in-situ* reflection high energy electron diffraction (RHEED) was used for real-time crystalline structure determination during the deposition. The chamber is pumped to ~1.0 x 10^{-7} Torr before IBAD deposition commenced. The MgO vapor flux was provided by e-beam evaporation with a deposition rate of 0.15 nm/s monitored by a quartz crystal oscillator near the substrate surface. A background oxygen partial pressure of 7.0 x 10^{-5} torr was maintained during Ar^+ ion pre-exposure and throughout the MgO growth process. XRD was used to examine the texture quality. The surface morphology of the bicrystal grain boundary region was measured by a KLA Tencor P-16 profiler. The out-of-plane texture of the templates was characterized by XRD rocking curves and the in-plane texture with XRD Φ scans. The surface morphology of the samples and substrates were studied by AFM.

DISCUSSION

Textured IBAD-MgO template on hard substrates

On all substrates we have studied, it was found that the texture quality of the IBAD-MgO textured template is sensitively dictated by substrate surface condition such as substrate surface roughness (R_a) as determined using AFM. R_a of less than 1 nm was found to be critical in bi-axial texture formation in IBAD-MgO films[10-12]. This requirement entails careful substrate preparation including electro-polishing metallic substrates so as to reach substrate roughness of R_a<1 nm. On non-metallic substrates, surface polishing processes need to be developed. Two such processes were developed in our work. One is to use ion beam pre-sputtering, which applies well to glass substrates of R_a typically on the order of several nm and the other, spin-on-glass (SOG) coating for much rougher substrates like ceraflex of R_a around tens to hundreds of nm. In the former, the same Ar^+ ion beam for IBAD-MgO texturing was employed and the R_a shows a continuous improvement from 2-3 nm to <1 nm after pre-sputtering for 10 minutes or slightly longer[14]. On the latter, a multilayer coating of the SOG was found efficient to systematically reduce R_a. In fact, with three coatings of SOG, R_a ~1 nm was achieved[16,17].

It should be realized that the R_a<1 nm is not the adequate condition for IBAD-MgO textured template to form. On commercial SiO_2/Si substrates, the textured MgO does not form directly although the R_a<1 nm is satisfied. Generally, depositing a buffer layer of typically 5-10 nm thick amorphous silicon nitride (Si_3N_4)[6,8-10,12] or yttrium oxide (Y_2O_3)[11] is necessary for high quality textured MgO to form on substrates even their R_a<1 nm is already in scope. Considering the requirement of substrate optical transmission for many optical device applications, we have explored pre-sputter the substrates before IBAD process and found this procedure plays a critical role in forming highly textured IBAD-MgO[15]. On one hand this is good news since IBAD-MgO can be formed without any buffers on glass and SiO_2/Si substrates, but on the other hand raises a question on what exactly happened during the pre-sputtering. It is speculated that these additional surface treatment may activate the surface by eliminating thin passive layers on the substrate surface. It should also be mentioned that surface smoothening using ion-beam pre-sputtering cannot be applied universally to all substrates. On ceraflex with SOG top layers, Y2O3 buffer layer was found necessary since ion beam pre-sputtering roughens the SOG surface and prevents IBAD-MgO textured template to form atop. Table 1 summarizes the texture quality of the IBAD-MgO textured templates on four different substrates presented in this work.

Table 1. Surface Roughness of Substrate and Texture Quality of IBAD-MgO

Substrate	Surface Roughness R_a of Substrate		Texture Quality of IBAD-MgO (with homo-epi MgO layer)	
	Original	Modified	In Plane (FWHM of Φ-scan)	Out of Plane (FWHM of ω-scan)
SiO_2/Si	≤ 1 nm	≤ 1 nm (Ar^+ pre-bombarded)	~ 6.5°	~ 2.0°
Glass	~ 2-3 nm	≤ 1 nm (Ar^+ pre-bombarded)	~ 6.5°	~ 2.0°
Ceraflex	~ 50-100 nm	≤ 1.5 nm (multilayer SOG smoothened) ≤ 1.5 nm (Y_2O_3 buffered)	~ 9.3°	~ 2.8°
PI/Si	≤ 1 nm	~ 1-2 nm (Y_2O_3 buffered)	~ 9°-10°	~ 3.0°

Textured IBAD-MgO template on soft substrates

Our work on development IBAD-MgO textured templates on soft substrates was motivated by need to integrate ferroelectric thin film devices with silicon complementary metal-oxide-semiconductor technology to enable commercially viable high density nonvolatile ferroelectric memories, ferroelectric IR detectors and other technologies[20-23]. Of particular interest for epitaxial growth on non-textured substrates is the fabrication of uncooled infrared (IR) ferroelectric sensors. A ferroelectric IR detector such as thin film ferroelectric device (TFFE) has a multilayer structure, including layers for bottom and top electrodes, the ferroelectric layer and a sacrificial layer which will later be removed to form a suspended structure[24]. The quality of the ferroelectric film, such as phase purity and crystalline orientation, is critical in the overall performance of the TFFE IR detector. Additionally, such detectors must be thermally isolated from the heat sink to increase sensitivity. A suspended microbridge structure can achieve high sensitivity because it prevents the heat on the detector active element escaping to the heat sink. However, since the barrier/sacrificial layer is often amorphous or polycrystalline, the whole ferroelectric stack grown on the sacrificial barrier, is at most polycrystalline. For instance, $Pb(ZrTi)O_3$ (PZT) ferroelectric film is promising for IR detection because of its large pyroelectric effect. PZT has a perovskite structure and is thus highly anisotropic. The best pyroelectric/ferroelectric effect is observed along the c-axis perpendicular to the perovskite planes. To maximize the pyrolectric effect, which in turn, affect the figure-of-merit for the TFFE detector, PZT must be grown in c-axis orientation. It is therefore of interest to explore the possibility of growing epitaxial ferroelectric layers on amorphous sacrificial layer using IBAD-MgO textured templates (see Figure 1).

Polyimides are high temperature engineering polymers which exhibit exceptional thermal stability, mechanical toughness, chemical resistance and low dielectric constant. Organic polyimides compose of long chains of the imide group. Polyimide films are frequently used as protective overcoat for semiconductors. They are used primarily to protect delicate thin films of metal and oxides on the chip surface from damage during handling. Polyimides may also serve as an interlayer dielectric in both semiconductors and thin film multichip modules[25-27]. Perhaps the biggest advantage of polyimides over conventional substrates is its excellent elastic properties. This is especially important in applications which require flexible substrates such as tapes and foils. Such properties make polyimide films attractive in thin film device fabrication.

Figure 1. A schematic diagram showing the insertion of a textured template material for the epitaxial growth of TFFE material on an amorphous polymer film as a sacrificial layer.

The effect of ion beam pre-exposure on polyimide films has been studied earlier[28]. It was observed that when the polyimide film surfaces were exposed to Ar^+ and O^+ ion beams, the PI surface developed large etch pits and grass like features. The roughening of the PI surface after ion pre-exposure was attributed to the inhomogenity of ion beam etch rates of the chains of the polyimide. Ar^+ ions have been observed to sputter the oxygen and nitrogen from the polyimide chains. The O^+ ions on the other hand interact both physically and chemically. Reactive oxygen species can also attach to the polyimide chain thereby weakening the carbon-carbon bonds. The polyimide chain has no long range crystalline structure but composed chains of ordered and disordered phases. When exposed to ion beam, the etch rate of the ordered phases is less than that for the disordered phases because of their higher density and more efficient bonding. This difference in the etch rate results in weakened bonds and can cause portions of the polymer chain to break, thereby increasing the roughness of the PI surface[28].

The ion beam roughening of the PI film surface was confirmed in our experiment as shown in Fig. 2. After 5 minutes of the ion beam pre-sputtering, the R_a increased from less than 1nm [Fig. 2(a)] to about 5nm [Fig. 2(b)]. To protect the polyimide surface from direct ion beam exposure while not adding additional layers to the IBAD MgO template, a 25 nm Y_2O_3 layer was first evaporated on the PI/Si as a buffer layer using e-beam evaporation and then removed partially via Ar^+ ion beam pre-exposure before deposition of the IBAD MgO templates. Highly textured MgO templates were obtained on Y_2O_3 buffered PI film with Ar^+ pre-exposure up to 5 minutes. Longer exposure time <10 minutes resulted in poor texture quality in MgO templates and no in-plane texturing was obtained due to complete removal of the Y_2O_3 layer. Fig. 2(c) shows the Y_2O_3 surface on polyimide/Si after 5 minutes of Ar^+ ion exposure. R_a ~ 1.5-2.0 nm was much improved over that on the bare PI surface after the same Ar⁺ ion exposure. At this exposure time, the remaining Y_2O_3 layer is about 5 nm. The best texture achieved in this study in homo-epi MgO layer is 9.0-10.0° in-plane via (220) MgO φ-scan and 3.0° out-of-plane via MgO (005) rocking curve. Fig. 2(d) shows the surface morphology of such as sample.

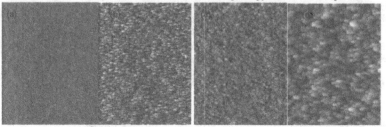

Figure 2. AFM images of surface morphology of (a) original PI film of R_a <1.0 nm ; (b) after ion beam pre-sputtering of 5 minutes. R_a~4-5 nm; (c) after 5 minutes ion beam pre-sputtering of 25 nm thick Y_2O_3 on PI film R_a~1.5-2 nm; and (d) 100 nm homo-epitaxial MgO deposited on 10 nm IBAD-MgO on sample in (c) with R_a ~2.0 nm.

Bi-crystal Textured IBAD-MgO templates

Bicrystals have been widely used to explore grain boundary effects and develop advanced devices. Grain boundaries can dramatically alter electron transport properties in many materials. The commercial bicrystal fabrication techniques use hot pressing or sintering at temperatures typically exceeding 1000 °C, which are high cost and low efficiency. Moreover, the conventional techniques are limited mostly for fabrication of bicrystals with one straight grain boundary. Although few reports have been made on tricrystals, it is generally difficult, if it is not impossible, to apply these techniques for complicated arrays of bicrystals with microscopic pixel dimensions.

The IBAD texturing approach has major advantage in generating complicated bicrystal arrays in combined with various lithography approaches. Basically, the in-plane and out-of-plane orientations of the template can be controlled by the orientation of the incident ion beam with respect to the substrates. This provides great flexibility in designing the template orientations along both the out-of-plane and in-plane directions. Since the texture formation can be completed at about 10 nm thickness in IBAD-MgO templates, the shadowing effect of the various lithographic masks can be much reduced. We have made a preliminary assessment of the IBAD texturing process for generation of bicrystals and bicrystal arrays recently[19]. Both metal masks and photoresist masks were applied in a two-step IBAD process. As shown in Fig. 3, the first IBAD-MgO textured template can be made using the same procedure described earlier in this paper. A mask will then be applied and a second IBAD-MgO textured template can be laid on the area not protected by the mask. In Fig. 3, a TEM grid with a hole dimension of 25 μm was used as the mask. In this specific case, the sample was rotated by 45 degree in the plane of the substrate with respect to the incident ion beam so the grain boundary angle of 45 degree is expected and confirmed in XRD. This result suggests that the IBAD texturing is a promising method for generating microscopic bicrystal arrays.

Grid size: 25 μm x 25 μm

Figure 3. Schematic description of IBAD-MgO bi-crystal template fabrication process using TEM grid as shadow mask. Left panel: after the first IBAD-MgO template was completed, a TEM grid was placed on the sample; middle panel: optical image of the TEM grid; and right panle: optical image of the bi-crystal array after the second IBAD-MgO template was laid on the "holes".

CONCLUSIONS

The successful generation of MgO templates on various surfaces may pave the way for the development of high quality multilayer devices on technologically interesting substrates that are amorphous or polycrystalline. A systematic research is necessary to reach such a goal through understanding of basic physics underlying the IBAD texturing process.

ACKNOWLEDGMENTS

RV was supported by MDA-ARO and JW and RTL acknowledge supports from DOE and AFOSR.

REFERENCES

1. 1. L.S. Yu, J.M.E. Harper, J. J. Cuomo and D. Smith: Alignment of thin films by glancing angle ion bombardment during deposition. *Appl. Phys. Lett.* **47**, (9) 932 (1985).
2. Y. Iijima, N. Tanabe, O. Kohno and Y. Ikeno: In-plane aligned $Yba_2Cu_3O_{7-x}$ thin films deposited on polycrystalline metallic substrates. *Appl. Phys. Lett.* **60**, (6) 769 (1992).
3. R.P. Reade, P. Berdahl, R.E. Russo and S.M. Garrison: Laser deposition of biaxially textured yttria-stabilized zirconia buffer layers on polycrystalline metallic alloys for high critical current Y-Ba-Cu-O thin films. *App. Phys. Lett.* **61**, (18) 2231 (1992).
4. M.P. Chudzik, R. Erck, M.T. Lanagan and C.R. Kannewurf: Processing dependence of biaxial texture in yttria stabilized zirconia by ion-beam-assisted deposition. *IEEE Trans. Appl. Supercon.* **9**, (2) 1490 (1999).
5. X.D. Wu, S.R. Foltyn, P.N. Arendt, J. Townsend, C. Adams, I.H. Campbell, P. Tiwari, Y. Coulter and J.E. Peterson: High current $YBa_2Cu_3O_{7-\delta}$ thick films on flexible nickel substrates with textured buffer layers. *Appl. Phys. Lett.* **65** (15) 1961 (1994).
6. R. Huhne, C. Beyer, B. Holzapfel, C.G. Oertel, L. Schultz and W. Skrotzki: Formation and destruction of cube texture in MgO films using ion beam assisted pulsed laser deposition. *J. Appl. Phys.* **90**, (2) 1035 (2001).

7. S.R. Foltyn, P.N. Arendt, Q.X. Jia, H. Wang, J.L. MacManus-Discoll, S. Kreiskott, R.F. De Paula, L. Stan, J. R. Groves and P.C. Dowden: Strongly coupled critical current density values achieved in $Y_1Ba_2Cu_3O_{7-\delta}$ coated conductors with near-single-crytsal texture. *Appl. Phys. Lett.* **82**, (25) 4519 (2003).

8. C.P. Wang, K.B. Do, M.R. Beasely, T.H. Geballe, and R.H. Hammond: Deposition of in-plane textured MgO on amorphous Si_3N_4 substrates by ion-beam-assisted deposition and comparisons with ion-beam-assisted yttria-stabilized-zirconia. *Appl. Phys. Lett.* **71**, (20) 2955 (1997).

9. R.T. Brewer and H.A. Atwater: Rapid biaxial texture development during nucleation of MgO thin films during ion beam-assisted deposition. *Appl. Phys. Lett.* **80**, (18) 3388 (2002).

10. R. T. Brewer, H.A. Atwater, J.R. Groves and P.N. Arendt: Reflection high-energy electron diffraction experimental analysis of polycrystalline MgO films with grain size and orientation distributions. *J. App. Phys.* **93**, (1) 205 (2003).

11. J.R. Groves, P.N. Arendt, H. Kung, S.R. Foltyn, R.F. DePaula, L.K. Emmert and J.G. Storer: Texture development in IBAD MgO films as a function of deposition thickness and rate. *IEEE Trans. Appl. Supercon.* **11**, (1) 2822 (2001).

12. A.T. Findikoglu, S. Kreiskott, P.M. te Riele and V. Matias: Role of beam divergence and ion-to-molecule flux ratio in ion-beam-assisted deposition texturing of MgO. *J. Mater. Res.* **19**, (2) 501 (2003).

13. J.R. Groves, P.N. Arendt, Q.X. Jia, S.R. Foltyn, R.F. De Paula, P.C. Dowden, L.R. Kinder, Y. Fan and E.J. Peterson: High critical current density PLD YBCO deposited on highly textured IBAD MgO buffer layers. *Ceramic Transactions* **104**, 219 (2000).

14. C.P. Wang: Ion beam modification of oxide thin film texture and its applications. *PhD Thesis* p.17, Stanford University (1999).

15. R. Vallejo and J.Z. Wu, "Ion Beam Assisted Deposition of textured Magnesium Oxide templates on un-buffered glass and silicon substrates", *J. Mat. Res.* **21**, 194 (2006)

16. Rongtao Lu, C. Varanasi, J. Burke, I. Maartense, P.N. Barnes and J. Z. Wu, "Textured IBAD-MgO Template on Non-Metallic Flexible Ceraflex for Epitaxial Growth of Perovskite Films", **invited** paper to MS&T Conference Proceeding, Cincinnati, OH, Oct. 16-20, 2006.

17. R.T. Lu, R. Vallejo and J.Z. Wu, "Development of textured MgO templates on nonmetallic flexible cereflex", Appl. Phys. Lett. **89**, 132505 (2006).

18. C.V. Varanasi, J. Burke, R. Lu, J. Wu, L. Brunke, L. Chuck, H.E. Smith, I. Maartense, P.N. Barnes, "Biaxially textured $YBa_2Cu_3O_{7\_x}$ films deposited on polycrystalline flexible yttria-stabilized zirconia ceramic substrates", Physica C (2008), doi:10.1016/j.physc.2008.05.258

19. R. Vallejo, R.T. Lu and J.Z. Wu, "Generating bi-crystal templates and arrays using ion beam assisted texturing", *Solid Thin Films* **517**, 609 (2008).

20. J.R. Roeder, I.S. Chen, P.C. Van Buskirk, H.R. Beratan and C.M. Hanson: Dielectric and pyroelectric properties of thin film PZT. *Proc. XI ISAF*, IEEE **217** (1998).

21. C.C. Yang, M.S. Chen, T.J. Hong, C.M. Wu, J.M. Wu and T.B. Wu: Preparation of (100)-oriented metallic $LaNiO_3$ thin films on Si substrates by radio frequency magnetron sputtering for the growth of textured $Pb(Zr_{0.53}Ti_{0.47})O_3$. *Appl. Phys. Lett.* **66**, (20) 2643 (1995).

22. M. Es-Souni, M. Kuhnke, S. Iakovlev, C.H. Solterbeck and A. Piorra: Self-poled $Pb(Zr,Ti)O_3$ films with improved pyroelectric properties via the use of $(La_{0.8}Sr_{0.2})MnO_3$/metal substrate heterostructures. *Appl. Phys. Lett.* **86**, 022907 (2005).

23. N.A. Basit, H.K. Kim and J. Blachere: Growth of highly oriented $Pb(Zr,Ti)O_3$ films on MgO-buffered oxidized Si substrates and its application to ferroelectric nonvolatile memory field-effect transistors. *Appl. Phys. Lett.* **73**, (26) 3941 (1998).
24. C.M. Hanson, "Hybrid pyroelectric-ferroelectric bolometer arrays", book chapter in *Semiconductors and semimetals*, **47**, 123 (1997).
25. M. Ree, K.J. Chen, D.P. Kirby, N. Katznellenbogen and D. Grischkowsky, " Anisotropic properties of high-temperature polyimide thin films: dielectric and thermal expansion behaviors", *J. App. Phys.* **72**, 2014 (1992).
26. M.I. Bessonov,M.M. Koton, V.V. kudyavstev and L.A. Laius, Polyimides-Thermally Stable Polymers. p.272, Consultants Bureau, New York (1987).
27. Y.N. Sazanov, "Applied Significance of Polyimides", *Russian Journ. of Appl. Chem.* **74**, 1253 (2001).
28. A. Ruoff, E.J. Kramer and C. Yu, Improvement of adhesion of copper on polyimide by reactive ion-beam etching", *IBM J. Res. Develop.* **32**, 626 (1988).

Mater. Res. Soc. Symp. Proc. Vol. 1150 © 2009 Materials Research Society   1150-RR03-02

Ion-beam assisted pulsed laser deposition of textured transition-metal nitride films

Ruben Hühne, Martin Kidszun, Konrad Güth, Franziska Thoss, Bernd Rellinghaus, Ludwig Schultz and Bernhard Holzapfel

IFW Dresden, Institute for Metallic Materials, P.O. Box 270116, D-01171 Dresden, Germany

ABSTRACT

Ion-beam assisted deposition (IBAD) offers the possibility to prepare thin textured films on amorphous or non-textured substrates. It was shown within the last decade that the ion beam influences the nucleation in material with a rocksalt structure leading to a strong cube texture already within the first 10 nanometres. Among these materials, transition metal nitrides exhibit interesting physical properties as superconductivity, metallic behaviour or superior hardness. Therefore, a reactive IBAD process was applied for the preparation of highly textured transition metal nitride layers using pulsed laser deposition of pure metals in combination with a nitrogen-containing ion beam. The results on the in-plane textured growth of TiN are promising for the development of conducting buffer layer architectures for YBCO coated conductors based on the IBAD approach. Furthermore, this approach was used to prepare other highly textured transition metal nitride thin films like NbN and ZrN.

INTRODUCTION

It was shown within the last decade that highly textured MgO films can be prepared on amorphous or nanocrystalline seed layers using ion beam assisted deposition (IBAD) [1-3]. The ion beam influences the nucleation leading to a strong cube texture within the first 10 nanometres. The <110> axis of the grown film is aligned parallel to the ion beam in these materials, if an ion incidence angle of 45° is used relative to the substrate normal. More recently it was found that other materials with a rocksalt structure, for example TiN or NbN, can be textured in a similar way [4,5]. This phenomenon has been explained by an anisotropy of either sputter yield or radiation damage, which favours the nucleation of grains with a defined orientation relationship towards the ion beam [6-8]. In most cases, the texture changes from the desired (100)[001] cube texture to texture components with a <100> direction parallel to the ion beam with increasing film thickness [3]. However, the cube textured layers can be stabilized to a higher film thickness using homoepitaxial growth without ion-beam assistance in order to use the films as templates for functional materials [1,4,9]. This paper summarizes our present work on the preparation of IBAD-textured nitride layers.

EXPERIMENT

Commercially available amorphous Si_3N_4/Si substrates were cut to a size of 10 mm x 10 mm and introduced into a standard PLD chamber in order to prepare the nitride layers in a reactive deposition. A Lambda Physics KrF excimer laser ($\lambda = 248$ nm) was applied to ablate pure Ti, Nb or Zr targets using laser energy densities of 2-8 J/cm^2 and laser pulse repetition rates of 1-10 Hz. An rf plasma source fed with a mixture of argon and nitrogen (ratio 1:1) was used as the assisting ion beam impinging the substrate surface at an angle of 45°. The mean ion beam energy was 800 eV with an ion flux density of about 50 µA/cm². The chamber pressure during deposition was about $1*10^{-3}$ mbar due to the gas flow from the ion gun. The thickness of the highly textured nucleation layer was increased using homoepitaxial growth of the transition metal nitrides at a substrate temperature of 700°C. In some cases, an additional 10 nm thin Au layer was epitaxially grown at 600°C in order to study the surface texture of the prepared nitride layers using X-ray diffraction. The texture formation was observed *in situ* using a STAIB Instruments reflection high energy electron diffraction (RHEED) system. Typically an electron energy of 30 keV and beam currents of about 50 µA were used under a grazing incidence angle of 0.5-1.5° to the substrate surface in order to examine the film growth. The diffraction pattern was recorded using a CCD camera. The grown films were characterised after deposition using X-ray diffraction (XRD) methods, which also allow the determination of the full-width at half-maximum (FWHM) of the *in-plane* and *out-of-plane* distribution. Additionally, scanning electron microscopy and atomic force microscopy (AFM) were used to investigate the surface structure and to measure the local roughness of the prepared layers. Selected cross sections have been prepared conventionally for TEM investigations.

RESULTS

Different transition metal nitride layers were prepared using reactive pulsed laser deposition with ion beam assistance. It should be noted that, in contrast to ion-beam assisted electron beam evaporation, the preparation of textured layers using ion-beam assisted pulsed laser deposition in general requires an elevated substrate temperature in order to achieve a strong cube texture [3,4]. The main focus of the work was on the preparation of highly textured TiN as a possible buffer layer for YBCO coated conductor applications. More recently, the investigations were extended to the preparation of textured superconducting nitrides as for example NbN or ZrN.

Preparation of IBAD TiN layers

RHEED investigations on the Si$_3$N$_4$ substrates revealed an amorphous surface structure, which was not destroyed during heating of the substrates up to temperatures above 400°C as shown for example in Fig. 1(a). A sharp nucleation texture was observed in TiN films within the first few nanometres using ion-beam assisted pulsed laser deposition. The typical pattern of a cube texture is shown in the RHEED image after deposition of about 5 nm at a substrate temperature of 400°C (Fig. 1(b)). In this case, the <100> direction of the crystal structure is parallel to the surface normal, whereas the ion beam is parallel to a <110> direction. The in-plane alignment was optimized modifying the laser energy densities on the target. This leads to different deposition rates and changes the ratio of the deposited atoms to the applied ions, which is know to influence the texturing severely [10]. The ion beam was switched off, when the diffraction pattern indicated a strong biaxial alignment in order to avoid the destruction the desired cube texture during further ion beam assistance. Instead, the thickness of the cube textured TiN layer was increased using a standard reactive pulsed laser deposition at a temperature of 700°C in a nitrogen-containing atmosphere [4]. The real-time RHEED pattern confirmed a homoepitaxial growth with sharper diffraction spots for an increasing film thickness as shown in Fig. 1(c) for a 90 nm thick TiN layer. The elongation of the spots perpendicular to the shadow edge is a clear indication for a smooth surface. The corresponding pole figure is shown in Fig. 1(d) having an *out-of-plane* FWHM of about 5° and an *in-plane* alignment of about 15°. The rms-roughness of such homoepitaxial grown TiN layers are typically between 1 and 2 nm (Fig. 2(b)). For comparison, the roughness of the substrate was measured to a value of about 0.2 nm.

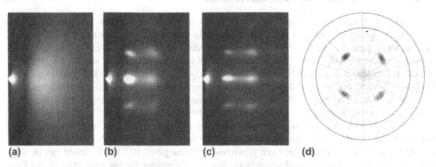

Figure 1. *In situ* RHEED pattern showing the surface texture of: (a) the amorphous Si_3N_4/Si substrate; (b) a 5 nm thick IBAD-TiN layer grown at 400°C; (c) a 90 nm thick TiN grown homoepitaxial at 700°C on the IBAD seed layer; (d) (220) X-ray pole figure of the 90 nm thick TiN layer

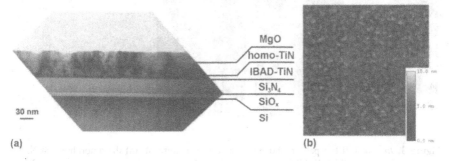

Figure 2. (a) TEM cross section of a textured 40 nm thick TiN layer with an additional 20 nm thick epitaxial MgO layer grown on Si/Si_3N_4; (b) AFM surface morphology of an about 90 nm thick textured TiN layer grown on Si/Si_3N_4

Cross sections of selected samples were prepared for TEM investigations in order to study the layer structure in more detail. One example is shown in figure 2(a). In this case, an additional MgO buffer layer was grown epitaxially on top of the cube textured $Si/Si_3N_4/IBAD$-

TiN/homoepitaxial-TiN architecture [9]. The TEM cross section indicates a smooth and dense TiN layer with a grain size below 30 nm.

It should be noted that the preparation of cube textured TiN was successfully transferred to technical substrates, i.e. to polished Hastelloy tapes with different amorphous seed layers [5,11]. In this case, an *in-plane* alignment down to 8° was achieved using optimized deposition conditions. The recent achievements in the development of a conducting buffer architecture based on such textured IBAD-TiN are summarized in a separate paper in this volume [12].

IBAD texturing of other transition metal nitrides

A similar cube textured nucleation as already described for TiN was observed during the ion-beam assisted deposition of other transition metal nitrides. Typical *in situ* RHEED pattern are shown in figure 3(a)-(c) for the preparation of cube textured NbN layers on Si/Si$_3$N$_4$ at a substrate temperature of 400°C. The diffraction pattern of a highly textured surface evolves from the diffuse pattern of the amorphous seed layer already after the first 5 nm (Fig. 3(b)). This cube nucleation texture is destroyed during further ion-beam assistance in a similar way as described for TiN or MgO previously [3,4]. However, the cube texture can be preserved to higher NbN thicknesses using homoepitaxial growth at a substrate temperature of 700°C. An *in-plane* FWHM value of about 14° was measured for an additional thin Au layer on top of the homoepitaxial grown NbN representing the surface texture of the nitride layer (Fig. 3(d)). It should be noted that different nitrogen pressures in the range of 10^{-2} to 10^{-3} mbar can be used during homoepitaxial growth and severely influence the superconducting properties of the final NbN layer. Superconducting transition temperatures up to 13 K were achieved in highly textured NbN layers, more details will be published separately.

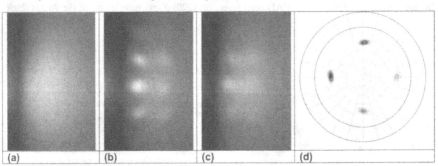

Figure 3. *In-situ* RHEED pattern showing the surface texture of: (a) the amorphous Si$_3$N$_4$/Si substrate; (b) a 5 nm thick IBAD-NbN layer grown at 400°C; (c) a homoepitaxially grown NbN with a thickness of about 50 nm prepared on the IBAD seed layer; (d) (111) X-ray pole figure of an additional Au layer.

The results of a cube textured ZrN layer prepared by ion-beam assisted deposition at a substrate temperature of 350°C are summarized in figure 4. *In situ* RHEED investigations reveal a similar texturing during nucleation as described for the other nitrides above (Fig. 4(b)). We attempted to preserve the cube texture to higher thicknesses using homoepitaxial growth at

700°C. Unfortunately, the intensity of the RHEED diffraction pattern became very weak during deposition, so that no clear statement could be made on the quality of the homoepitaxial growth. However, a broad cube texture is still visible in the XRD pole figure of the final ZrN layer (Fig. 4(c)) with some additional texture components indicating that the cube texture was partially preserved in the growth process. The IBAD step as well as the epitaxial growth needs to be optimized further in order to achieve single textured layers with a smaller *in-plane* FWHM. Nevertheless, it was shown for the first time that ion-beam assisted deposition is a suitable method to prepare highly textured ZrN layers.

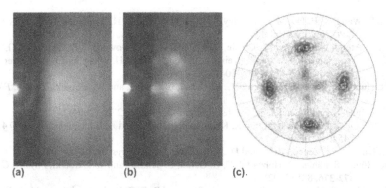

(a) (b) (c)

Figure 4. *In situ* RHEED pattern showing the surface texture of: (a) the amorphous Si_3N_4/Si substrate; (b) a 5 nm thick IBAD-ZrN layer grown at 350°C; (c) (111) X-ray pole figure of the ZrN layer after homoepitaxial growth at 700°C.

CONCLUSIONS

We have shown that different transition metal nitride layers with a NaCl structure can be textured during nucleation using reactive ion-beam assisted pulsed laser deposition. In-situ RHEED investigations revealed that in all cases the <110> axis of the grown film is aligned parallel to the assisting ion beam, whereas the <100> axis is parallel to the substrate normal. This observation strongly supports the assumption that in principle all metal nitrides having a rocksalt structure might be textured in a similar way during nucleation as originally shown for MgO. The observed (100)[001] cube texture is only stable in layers below 10 nm. Therefore, homoepitaxial growth at higher temperatures was used to preserve the strong nucleation texture to higher thicknesses. It was shown that textured TiN as well as NbN films can be prepared with an in-plane alignment below 15° using this approach. TEM cross sections revealed dense nitride layers with smooth interfaces. A similar textured nucleation was observed for the reactive deposition of ZrN; however, the homoepitaxial growth needs to be optimized in the future.

ACKNOWLEDGMENTS

The work and results reported in this letter were partially obtained with research funding from the European Community under the Sixth Framework Programme Contract No. 516858: HIPERCHEM.

REFERENCES

1. C. P. Wang, K. B. Do, M. R. Beasley, T. H. Geballe, and R. H. Hammond, *Appl. Phys. Lett.* **71**, 2955 (1997).
2. P. N. Arendt, S. R. Foltyn, L. Civale, R. F. DePaula, P. C. Dowden, J. R. Groves, T. G. Holesinger, Q. X. Jia, S. Kreiskott, L. Stan, I. Usov, H. Wang, and J. Y. Coulter, *Physica C* **412-414**, 795 (2004).
3. R. Hühne, C. Beyer, B. Holzapfel, C. G. Oertel, L. Schultz, and W. Skrotzki, *J. Appl. Phys.* **90**, 1035 (2001).
4. R. Hühne, S. Fähler, and B. Holzapfel, *Appl. Phys. Lett.* **85**, 2744 (2004).
5. R. Hühne, K. Güth, M. Kidszun, R. Kaltofen, L. Schultz, and B. Holzapfel, *J. Phys. D* **41**, 245404 (2008).
6. L. Dong, L. A. Zepeda-Ruiz, and D. J. Srolovitz, *J. Appl. Phys.* **89**, 4105 (2001).
7. R. Hühne, S. Fähler, B. Holzapfel, C. G. Oertel, L. Schultz, and W. Skrotzki, *Physica C* **372-376**, 825 (2002).
8. I. O. Usov, P. N. Arendt, J. R. Groves, L. Stan, and R. F. DePaula, *Nucl. Instr. Meth. Phys. B* **240**, 661 (2005).
9. R. Hühne, S. Fähler, L. Schultz, and B. Holzapfel, *Physica C* **426-431**, 893 (2005).
10. A. T. Findikoglu, S. Kreiskott, P. M. te Riele, and V. Matias, *J. Mater. Res.* **19**, 501 (2004).
11. K. Güth, R. Hühne, V. Matias, J. Rowley, T. Thersleff, L. Schultz, and B. Holzapfel, *IEEE Trans. Appl. Supercond.* **19**, accepted (2009).
12. R. Hühne, K. Güth, M. Kidszun, R. Kaltofen, V. Matias, J. Rowley, L. Schultz, and B. Holzapfel, *MRS Proceedings 1150E (this volume)*, RR04-01 (2008).

Mater. Res. Soc. Symp. Proc. Vol. 1150 © 2009 Materials Research Society   1150-RR03-04

Formation of Biaxial Crystalline Texture by Oblique Ion Bombardment

Paul Berdahl, Ronald P. Reade, and Richard E. Russo
Environmental Energy Technologies Division, Lawrence Berkeley National Laboratory,
Berkeley, CA 94720

ABSTRACT

The historical background for Ion TEXturing (ITEX) is briefly reviewed, including Ion Beam Assisted Deposition (IBAD). Also of interest is the orientation of metal films with high-energy ions and the formation of crystallites in SiO_2 with low-energy neutral atom bombardment. We also speculate about future ITEX processes, especially texturing of amorphous Ge, Si and C.

The basic concept of ITEX is to obliquely bombard an amorphous material, to produce fully oriented crystallization at the surface. As a concrete example, ion texturing of yttria-stabilized zirconia (YSZ) can rapidly produce an out-of-plane (211) crystalline texture when oxygen-deficient amorphous YSZ is bombarded obliquely with Ar ions. The in-plane texture is (111) parallel, and (110) transverse, to the azimuth of the ion beam. Rapid two-dimensional Ostwald ripening accompanies production of the texture; crystallite diameter d exceeds 100 nm in 5 min. Theoretical arguments suggest that d is proportional to the time t as $t^{3/7}$ and the crystallite orientation angle distribution widths are proportional to $t^{-2/7}$. When the temperature is subsequently raised to 300 °C, the ion-induced surface texture grows down into the bulk of the film by epitaxial oxidation.

INTRODUCTION

For many applications, it is desirable to be able to form a highly textured film (near-single-crystal) on technical substrates such as glass and steel. Such applications include high-temperature superconducting tapes, photovoltaics, flat panel displays, etc. The textured film forms a template for subsequent epitaxial growth of another film that would otherwise require a single-crystal substrate. Also, the textured film itself can be the active material. One technique for producing such textured films is called ITEX (Ion TEXturing) [1-3], the subject of this paper. In an ITEX process, a precursor film is deposited first. The precursor film should be amorphous (or perhaps nanocrystalline), and is subsequently bombarded with a low-energy (< 1000 eV) oblique ion beam that causes it to crystallize. The process steps can potentially be rapid and inexpensive, but still result in a fully oriented polycrystalline film with only small-angle grain boundaries. So far, work is focused on cubic zirconia (YSZ, yttria-stabilized zirconia), but other materials systems are likely to show similar texturing phenomena.

The plan of this presentation has three parts. First, we outline some of the history of the formation of artificially induced grain alignment in thin films, with emphasis on the ion beam methods that led up to the recent ITEX research. Second, we describe the most studied ITEX process, the YSZ (211) process, as a specific example. Finally, we speculate on some of the other materials systems that merit investigation, especially the semiconductors Ge, Si, and C.

HISTORICAL BACKGROUND

Much of the recent research on artificial grain alignment in thin films has focused on substrates for high-temperature superconducting tapes. The three dominant methods are IBAD (Ion Beam Assisted Deposition), RABiTS (Rolling Assisted Bi-axial Textured Substrates), and ISD (Inclined Substrate Deposition). Of primary interest here is the IBAD work. [See other papers in this volume for suitable references on RABiTS and ISD.] However, a word about older ISD work is in order here. A recent, thorough review of MgO ISD work has been recently presented by Xu *et al*. [4]. These authors attribute the origin of ISD to work on magnetic films by D. O. Smith in 1959. However, there are some older papers. One such paper is in the first issue of the journal *Physica* in 1934 by Burgers and Dippel [5]. These authors studied evaporation of CaF_2 films, and found by electron scattering that (111) crystallites were normal to the film in the center of the deposition zone but tilted toward the source if the deposition was oblique. Modern ISD work on CaF_2 may be found in Ref. [6].

In the 1980's several groups examined how thin metal films could be oriented by an Ar^+ or other ion beam of 10 to 50 keV. Typically the ion beam was incident normally, so the texture achieved was of the fiber type with preferred normal orientation and random in-plane texture. At the high energy employed the entire thin film could be reoriented. If fcc films such as Ag, Au, and Cu were bombarded, the (110) fiber texture was produced. For the fcc structure, the nearest neighbor direction is (110), which is likewise the "channeling" direction. The work by Dobrev [7] is a good example of the work on metal films, and the explanation of the orientation by channeling is not in doubt. The mechanism(s) for texture development by lower energy ions in IBAD and ITEX films is more obscure.

In a paper published in 1995 [8], T. Mizutani reported a number of experiments on the effects of ion and neutral-atom beams on the amorphous phase of silica, SiO_2. He found that bombardment with 400 eV ions caused more sputtering than bombardment with the corresponding neutral atoms. Bombardment of metal (Cu) by ions and atoms causes equal amounts of sputtering, but the semiconductor Si and the insulating SiO_2 phases were more strongly affected by ions compared with atoms. He also observed preferential sputtering of oxygen, so that the silica phase could become oxygen deficient. Most interestingly, bombardment with Ne^0 and Kr^0 atoms (350 eV, 10^{17} cm^{-2}) can cause formation of quartz crystallites in the top 2 nm of an SiO_2 film. These crystallites were characterized by high-resolution TEM and RHEED. Bombardment with Kr^0 produced the α-crystobalite phase, while the Ne^0 beam produced the α-quartz phase. These crystallites were found in the top part of a 5 nm thick layer that was denser than the bulk of the SiO_2 film. Bombardment with Ar^0, Ar^+, Ne^+, and Kr^+ did not cause the formation of quartz crystallites. Thus it is clear that creation of crystallites in the surface region of an amorphous SiO_2 film can be caused by neutral atom bombardment, but specific favorable conditions are required.

The Berkeley Lab effort to make high-Jc high-Tc films on metal substrates

In May of 1988 a new U. S. Dept. of Energy program was initiated to develop the then-new high temperature superconductors for applications to electric power systems. Three of the national laboratories started major programs (Los Alamos, Oak Ridge, and Argonne), while several other laboratories, including Lawrence Berkeley, initiated smaller programs, with funding levels about one tenth the size of each of the big three. The Berkeley Lab program was specifically focused on fabrication of the needed conductors by deposition of thin superconducting films on flexible metal substrates. At the time, the foremost method for conductor fabrication was to place superconducting powders in metal tubes, to draw them out into wires, followed by heat treatments. The successful wires that resulted, "first generation wires," were based on the bismuth families of high-Tc conductors. In contrast, Berkeley's effort to use thin film technology and metal substrates to fabricate tape conductors was an innovative but high-risk effort.

In August of 1990, the results of the annual DOE Peer Review of the Berkeley Lab effort were as follows: "Our review committee (*as was last year's*) was pessimistic about the prospects for HTS [high temperature superconductor] thin films being successful as high-current conductors." Meanwhile, it had become clear that if the important material $YBa_2Cu_3O_7$ was to be used, the grains of the polycrystalline material had to be well aligned, with only small-angle grain boundaries. There were only a very limited number of possibilities to accomplish this goal. Grapho-epitaxy was one possibility, "lucky texture" another, and a growth-assisting oblique ion beam, a third possibility. We borrowed an ion source from another group and fabricated YSZ buffer layers by pulsed laser deposition (PLD) on a nickel superalloy substrate using the oblique IBAD method. PLD was then used to fabricate an epitaxial $YBa_2Cu_3O_7$ layer on the YSZ template. In June of 1992, we submitted a paper [9] demonstrating a critical current density of 6×10^5 A cm^{-2}. This was a world record at the time, and a factor of 50 improvement compared to films without in-plane orientation.

Our success in the face of pessimistic assessments by review committees just illustrates the well-known phenomenon that the consensus of a group of experts is not a good basis for innovation.

The status of coated conductors in 1992

Superconducting tape conductors made by thin film deposition on metal substrates are now known as "coated conductors" or "second generation conductors." However, in 1992 these materials were still rather novel. Iijima's group at Fujikura in Japan had fabricated similar structures in 1991 and 1992 [10,11], using sputtering rather than PLD for the IBAD YSZ layer, achieving a critical current density of 2.5×10^5 A cm^{-2}. Thus, based on the Berkeley and Fujikura results, it was at that point completely clear that coated conductors were technically feasible. Future applicability of the technology was "just" a question of cost and performance [12]. In the years that followed the main DOE program on high temperature superconductors shifted emphasis from the first to the second generation conductors, and now in recent years several companies can manufacture hundred-meter lengths of high-performance tape conductors. In 2007 Foltyn *et al.* published an excellent review [13] of the progress since 1992. It remains to be seen how low the cost of the conductors can become. If the cost can be driven down to compete with copper on an ampere-by-ampere basis, applications will be widespread indeed.

Limitations of the YSZ IBAD process

For the post-1992 efforts to make coated conductors more manufacturable, the YSZ IBAD processes were seen as too slow. The YSZ film thickness needed to be more than 0.5 micrometers, due to the gradual evolution of texture as the film grew. The cubic unit cell of YSZ is only 0.5 nm on edge; a thickness of more than 1,000 unit cells was required. In contrast, a layer of only two or three unit cells would be suitable for epitaxial growth of the superconducting layer if a more rapid texturing strategy could be developed. At Stanford, an MgO IBAD process was invented [14] that is very fast, and which is now in widespread use. At Berkeley, we focused on the new approach we termed ITEX. It likewise can produce a textured crystalline layer at high speed, but has not yet evolved beyond our own laboratory.

ION TEXTURING (ITEX) AND THE SPECIFIC EXAMPLE OF THE YSZ (211) PROCESS

The concept of ITEX can be thought of as simply IBAD without concurrent film deposition. A precursor film is synthesized first and, subsequently, the surface region of the film forms fully oriented crystallites during a brief, oblique ion bombardment. The precursor film needs to be amorphous or nanocrystalline – with stored potential energy relative to a crystalline structure – so that it "wants" to crystallize during the ion bombardment. Our initial ITEX work [1-3] is all based on the YSZ material system. The best studied process is a mostly room temperature process that results in a texture with the (211) orientation up, the (111) in-plane orientation parallel to the azimuth of the ion beam, and the (110) orientation in the third orthogonal (transverse) direction. The conditions for this process are listed in Table 1.

Table 1. Parameters for the YSZ (211) ITEX process.

YSZ (211) ITEX process conditions	
Amorphous YSZ layer synthesis	Reactive magnetron sputtering
Oxygen stoichiometry	Precursor layer is ~ 50% oxygen deficient
Ion bombardment	300 eV Ar^+ ions, 400 μA cm^{-2}, 55° from normal
Pressure during bombardment	0.4 mtorr Ar, 0.4 mtorr O_2
Temperature during bombardment	Room temperature; temp. raised at end to epitaxially oxidize film

D. Buczek at American Superconductor synthesized YSZ films using a metallic target and various oxygen pressures, finding that with ~ 50% oxygen deficiency an amorphous film could be formed [2]. During ITEX processing, the background gas contains oxygen so that the bombarded surface region is fully oxygenated, assisting crystallization. As with Mizutani's work [8] on SiO_2, we expect that any crystallites present are in roughly the top 2 nm of the film. However, in our case RHEED scattering does not show evidence for crystallites. The surface region may be disordered (proto-crystalline). Under the conditions used, the sputtering rate is approximately 0.05 nm s^{-1}. Thus, if the ion-modified zone is 2 nm thick, a kind of dynamic equilibrium is established in ~ 40 s. Bombardment times ranged from about 2 to 30 min, with even 2 min. being adequate for texturing. Toward the end of the bombardment the temperature is raised, permitting oxygen to rapidly diffuse through the top crystalline YSZ layer, epitaxially forming a transparent crystalline (211) layer. X-ray pole figures then document the orientation of the layer, with similar results also obtained by the analysis of an epitaxial CeO_2 layer grown on the (211) YSZ template [2]. The (211) layer may be thought of as a tilted structure. If one starts with a (001) film, and rotates the crystallites toward the ion beam about 35 deg, about a (110) in-plane axis, one has the (211) structure. The ion beam is not parallel to a low-index direction. Thus the orientation mechanism is not ion channeling. It also seems unlikely to be selective sputtering. The rapid grain coarsening, to be discussed next, appears to be due to ion-induced solid-state recrystallization, and this same mechanism is likely to be the cause of crystallite orientation. Favorably oriented crystallites resist ion damage and consume adjacent crystallites, enabled by the high ion-assisted grain boundary mobility.

Figure 1 shows images of the ion beam induced grain growth. These samples were not subjected to heating after bombardment and maintained the metallic appearance of the precursor films.

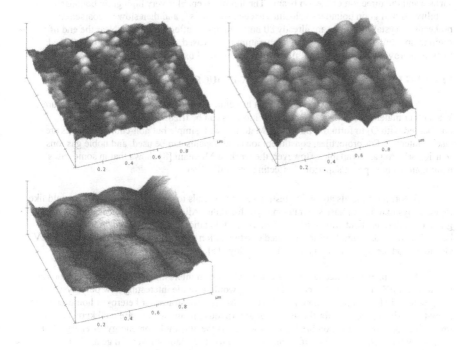

Figure 1. Atomic force microscope topographs of 3 partly-processed (211) YSZ ITEX films. Each image is 1 μm x 1 μm with vertical scale 30 nm. Linear features aligned with left edges are artifacts due to rolling marks in the underlying metal alloy substrates. Film grains are delineated by grain-boundary grooves. Grain size increases with bombardment time: 2 min., 5 min., and 7.5 min.

We have developed a simple theory for the grain coarsening process [3]. It assumes that the grain orientations are random at nucleation and that the "quality factor" that determines the rate by which the "more-fit" grains consume the "less-fit" grains is an analytic (smooth) function of the 3 Euler angles that describe grain orientation. In the asymptotic regime in which the grains are large compared to the original nuclei, the time dependence of the grain size scales like $d \sim t^{\beta}$, with $\beta = 3/7 = 0.43$. The 3 angular widths of the grain orientation angle distribution scale like $t^{-2/7}$. Experimentally, we estimated that $\beta = 0.5 \pm 0.2$, in good agreement with the theory.

The YSZ ITEX (211) process may be summarized as follows. When the precursor film is bombarded, widespread crystallite nucleation occurs in less than 1 second. These thin ("2-D") crystallites impinge on each other laterally and then consume each other in a "survival of the fittest contest." Fitness is determined by the orientation of the crystallites relative to the free surface and the direction of the ion beam. The ion beam enables very high grain boundary mobility. Initial grain boundary velocities exceed ~10 nm s^{-1}, and then slow as coarsening proceeds. Crystallite domes are already 80 nm in diameter after 2 minutes. Near the end of room temperature bombardment, the temperature is increased, which causes the ion-induced texture to grow down into the bulk of the film by epitaxial oxidation.

FINDING FUTURE PROCESSES FOR ION-INDUCED TEXTURE

Most of the current texturing research has employed metal oxides with cubic structures. What other materials can be ion textured? Fluorides can be tried. Perhaps they can be bombarded with O_2 to form oxyfluorides. Materials with simple but non-cubic structures are also of interest. For projectiles, constituent ions and dopants can be used, and noble gas ions (particularly Ar) are available. However, the work of Mizutani [8] shows that in some cases neutral atoms may prove superior in effecting crystallization.

Amorphous metals are an interesting set of materials to consider. Ideally, one would like to use alloys that don't phase separate on crystallization. Alloys used for optical storage are a group of materials about which the amorphous/crystallization phase transformation has already been studied. One example of ion-induced crystallization in an amorphous metal is that 30 keV Ga^{+} ions produce oriented Ni crystals in Ni-P alloy [15].

The elemental semiconductors – Ge, Si, C – are an interesting class of materials for ion texturing (ITEX or IBAD). Successful texturing would provide interesting new polycrystalline films with only low-angle grain boundaries and also new substrates for hetero- or homo-epitaxial growth. Various processes for the growth of amorphous precursor films are well known. An important issue for the ion bombardment, however, is the allowable ion energy, as energies of only a few hundred eV may be sufficient to amorphize a crystalline layer. In general, the use of elevated temperature is likely to be helpful, because it can push the amorphous phase closer to crystallization, and it can help to anneal out ion-induced defects.

For germanium, epitaxial growth at 400 °C can be enhanced by a 200 eV Xe^{+} beam [16]. Thus ion energies up to at least 200 eV Xe^{+} should be acceptable in a texturing process. In very brief experiments at Berkeley Lab [17], we found that a-Ge films that crystallized at 450 °C without ion bombardment (observed by RHEED) could be made to crystallize at 200 °C in the presence of a low-energy Ar^{+} beam. In-plane texture, however, was not achieved.

For silicon, one question that arises is: would hydrogen doping be helpful or not? [Hydrogen is likely to affect the crystallization process and can also passivate some silicon defects.] This question will not be answered here, but we point out that control over the chamber atmosphere and surface chemical reactions are important in most non-UHV (ultra high vacuum) processes. The ion beam can provide some surface cleaning. On the other hand, for example, in the YSZ ITEX and IBAD processes, the availability of oxygen is important to replace oxygen lost by preferential sputtering. One study [18] on silicon shows that 30 eV Ar^+ bombardment enables low-temperature (< 450 °C) growth of polycrystalline Si. Thus 30 eV bombardment in a Si ITEX process would be acceptable. The utility of higher energy ions is unclear.

An ion texturing process for carbon has been developed [19,20] for use in manufacture of liquid crystal displays (LCDs). The process first deposits an amorphous carbon film, and then bombards the film with oblique Ar^+ ions (45°, 200 eV, $4x10^{15}$ cm^{-2}). The textured layer thus formed can orient the liquid crystal molecules in LCDs. This process replaces the rubbing of a polymer film with cloth. The interpretation of the texturing mechanism is different from what we have described for YSZ ITEX. The Ar^+ ions break the carbon-carbon bonds at the surface, and the probability of breaking of the bonds is a function of bond orientation. Therefore, after bombardment the remaining bonds have an anisotropic in-plane orientation distribution. The rather large LCD molecules can then be aligned in registry with the carbon-carbon bonds. The optimum ion dose is small; each surface bond is struck only a small number of times.

SUMMARY

We have reviewed some of the historical background on ion texturing, including the original IBAD work on YSZ buffer layers that made it possible to fabricate the first high-current high-temperature superconducting tapes. More recent work on ITEX (Ion TEXturing) of YSZ was also discussed. The most-studied process results in a (211) fiber texture with in-plane orientation (111) parallel to the azimuth of the ion beam, and (110) transverse. This texture is accompanied by very rapid grain coarsening. The high mobility of the grain boundaries is caused by ion bombardment. Finally, we speculated on other materials that may be amenable to ion texturing, including fluorides, amorphous metal alloys, and the elemental semiconductors, Ge, Si, and C.

ACKNOWLEDGMENTS

The details of the YSZ (211) ion texturing process were investigated in collaboration with colleagues at American Superconductor (see Ref.[2]).

Lawrence Berkeley National Laboratory is operated by the University of California for the U. S. Department of Energy under contract DE-AC02-05CH11231.

REFERENCES

1. R. P. Reade, P. Berdahl, and R. E. Russo, *Appl. Phys. Lett. 80*, 1352 (2002).
2. P. Berdahl, R. P. Reade, J. Liu, R. E. Russo, L. Fritzemeier, D. Buczek, and U. Schoop, *Appl. Phys. Lett. 82*, 343 (2003).

3. P. Berdahl, R. P. Reade, and R. E. Russo, *J. Appl. Phys. 97*, 103511 (2005).
4. Y. Xu, C. H. Lei, B. Ma, H. Evans, H. Efstathiadis, M. Rane, M. Massey, U. Balachandran, and R. Bhattacharya, *Supercond. Sci. & Technol. 19*, 835 (2006).
5. W. G. Burgers and C. J. Dippel, *Physica 1*, 549 (1934).
6. H. –F Li, T. Parker, F. Tang, C. –G. Wang, T. –M. Lu, and S. Lee, *J. Crystal Growth 310*, 3610 (2008).
7. D. Dobrev, *Thin Solid Films 92*, 41 (1982).
8. T. Mizutani, *J. Non-Crys. Sol. 181*, 123 (1995).
9. R. P. Reade, P. Berdahl, S. M. Garrison, and R. E. Russo, *Appl. Phys. Lett. 61*, 2231 (1992).
10. Y. Iijima, N. Tanabe, O. Kohno, and T. Ikeno, *Physica C 185-189*, 1959 (1991).
11. Y. Iijima, N. Tanabe, O. Kohno, and T. Ikeno, *Appl. Phys. Lett. 60*, 769 (1992).
12. P. Berdahl, R. E. Russo, and R. P. Reade, *Chem. & Eng. News 73*, #26, 5 (June 26, 1995).
13. S. R. Foltyn, L. Civale, J. L. Macmanus-Driscoll, Q. X. Jia, B. Maiorov, H. Wang, and M. Maley, *Nature Materials 6*, 631 (2007).
14. C. P. Wang, K. B. Do, M. R. Beasley, T. H. Geballe, and R. H. Hammond, *Appl. Phys. Lett. 71*, 2955 (1997).
15. R. Tarumi, K. Takashima, and Y. Higo, *J. Appl. Phys. 94*, 6108 (2003).
16. E. Chason, P. Bedrossian, K. M. Horn, J. Y. Tsao, and S. T. Picraux, *Appl. Phys. Lett. 57*, 1793 (1990).
17. R. P. Reade and P. Berdahl, unpublished, (2003).
18. J. E. Gerbi and J. R. Abelson, *J. Appl. Phys. 101*, 063508 (2007).
19. P. Chauhari, J. Lacey, J. Doyle, E. Galligan, S.-C. A. Lien, A. Callegari, G. Hougham, N. D. Lang, P. S. Andry, R. John, K.-H. Yang, M. Lu, C. Cai, J. Speidell, S. Purushothaman, J. Ritsko, M. Samant, J. Stohr, Y. Nakagawa, Y. Katoh, Y. Saitoh, K. Sakai, H. Satoh, S. Odahara, H. Nakano, J. Nakagaki, and Y. Shiota, *Nature 411*, 56 (2001).
20. J. Stohr, M. G. Samant, J. Luning, A. C. Callegari, P. Chaudhari, J. P. Doyle, J. A. Lacey, S. A. Lien, S. Purushothaman, and J. L. Speidell, *Science 292*, 2299 (2001)

Mater. Res. Soc. Symp. Proc. Vol. 1150 © 2009 Materials Research Society   1150-RR03-05

Aligned-Crystalline Si Films on Non-Single-Crystalline Substrates

Alp T. Findikoglu[1], Terry G. Holesinger[2], Alyson C. Niemeyer[3], Vladimir Matias[1], and Ozan Ugurlu[4]

[1]MPA-STC, Los Alamos National Laboratory, MS T004, Los Alamos, NM 87545, U.S.A.

[2]MPA-STC, Los Alamos National Laboratory, MS K763, Los Alamos, NM 87545, U.S.A.

[3]T-1, Los Alamos National Laboratory, MS B268, Los Alamos, NM 87545, U.S.A.

[4]Characterization Facility, Institute of Technology, 55 Shepherd Laboratories, Minneapolis, MN 55455, U.S.A.

ABSTRACT

We summarize recent progress in growth and characterization of aligned-crystalline silicon (ACSi) films on polycrystalline metal and amorphous glass substrates. The ACSi deposition process uses, as a key technique, ion-beam-assisted deposition (IBAD) texturing on a non-single-crystalline substrate to achieve a *biaxially-oriented* (i.e., with preferred out-of-plane and in-plane crystallographic orientations) IBAD seed layer, upon which homo- and hetero-epitaxial buffer layers and hetero-epitaxial silicon (i.e., ACSi) films with good electronic properties can be grown. We have demonstrated the versatility of our approach by preparing ACSi films on customized architectures, including fully insulating and transparent IBAD layer and buffer layers based on oxides on glass and flexible metal tape, and conducting and reflective IBAD layer and buffer layers based on nitrides on flexible metal tape. Optimized 0.4-μm-thick ACSi films demonstrate out-of-plane and in-plane mosaic spreads of 0.8° and 1.3°, respectively, and a room-temperature Hall mobility of ~90 $cm^2/V.s$ (~50% of what is achievable with epitaxial single-crystalline Si films, and ~1000 times that of amorphous Si films) for a p-type doping concentration of ~$4x10^{16}$ cm^{-3}. By using various experimental techniques, we have confirmed the underlying crystalline order and the superior electrical characteristics of low-angle (<5°) grain boundaries in ACSi films. Forming gas anneal experiments indicate that Si films with low-angle grain boundaries do not need to be passivated to demonstrate improved majority carrier transport properties. Measurements on metal-insulator-semiconductor structures using ACSi films yield near-electronic-grade surface properties and low surface defect densities in the ACSi films. A prototype n+/p/p+–type diode fabricated using a 4.2-μm-thick ACSi film shows minority carrier lifetime of ~3 μs, an estimated diffusion length of ~30 μm in the p-Si layer with a doping concentration of $5x10^{16}$ cm^{-3}, and external quantum efficiency of ~80% at 450 nm with the addition of an MgO film anti-reflector.

INTRODUCTION

The motivation for using IBAD texturing for the preparation of ACSi films is the possibility to combine the high performance aspects of bulk single-crystalline Si with the cost advantages of Si films on amorphous or polycrystalline foreign (technical) substrates. Figure 1 illustrates our approach in obtaining ACSi films on a foreign substrate with the use of a functional buffer layer stack that includes an IBAD-textured layer (see Figure 1). The important attributes of the foreign substrate are: low cost, smooth surface (with rms surface roughness ≤10 nm), process compatibility with buffer layer and Si film growth, and additional application-specific functionality, such as flexibility, transparency, conductivity, low-weight, large-area, etc. Within the functional buffer, the IBAD textured layer, which is only 3-to-5 nm thick, is the key component that allows transition from no/low crystalline order to high crystalline order. The components of the functional buffer layer are chosen in such a way that, in addition to providing a robust crystalline template at the surface for hetero-epitaxial growth of ACSi films, they also add functionality in the bulk, such as high conductivity or transparency. The ACSi film that is grown epitaxially on the functional buffer layer needs to possess good crystallinity, and good electronic properties, such as high carrier mobility and high carrier lifetime, as well as application-specific properties, such as a well-defined doping profile within the film.

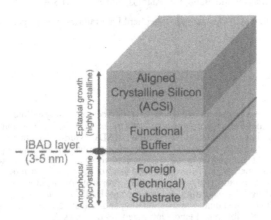

Figure 1. Schematic architecture of ACSi film on functional buffer on foreign substrate.

EXPERIMENTAL DETAILS

In the IBAD texturing process,[1,2] an off-normal ion beam establishes biaxially-oriented grains of a growing film of certain materials on (a non-single-crystalline film on) a non-single-crystalline substrate, thus establishing a highly crystalline template for the epitaxial growth of subsequent layers. The details of the IBAD texturing process are still not well understood,[1,3,4] although some computational and analytical modeling work has been done in the area of ion-atom interactions and their effect on texture formation.[5,6]

We have previously reported our work on the hetero-epitaxial growth of ACSi films on metal tapes[7,8,9,10] and glass plates[11] using functional buffer layers that include IBAD layers. In this paper, we summarize our previous work, and give comparative examples of different film architectures that we have recently prepared. Various alternative approaches to obtain crystalline Si on non-single-crystalline substrates have also been extensively explored by other research groups.[12,13,14,15]

All deposited layers (except for the homo-epitaxial growth of TiN by sputtering) reported in this paper were electron-beam evaporated in situ on either 1-cm-wide, 5-mil-thick continuous metal tapes, or 0.5-mm-thick glass substrates that were cut in 1 cm x 1 cm squares. We used two types of commercially-obtained metal tapes for this study: unpolished, cold-rolled Stainless Steel

305 with rms surface roughness of ~10-40 nm; and chemically-polished Hastelloy C276 Nickel-alloy with rms surface roughness of ~1-2 nm. The glass substrates were commercially-obtained, fusion-drawn, boro-alumino-silicate glass, with the thermal expansion coefficient closely matched to that of Si, and with a softening point ($10^{7.6}$ poises) of 985°C.

For the majority of samples reported here, we first deposited a roughly 5-nm-thick amorphous Si-N (Si-O) film at room temperature as a nucleation layer on the metal tapes (glass substrates). On this nucleation layer, we deposited a 3-to-5-nm-thick IBAD-textured TiN layer (MgO layer) with a 750-eV Ar^+ assist beam at ambient temperature, with ion/atom ratio of about 1, and a 45° angle between the ion beam and the substrate normal to form a biaxially-oriented crystalline template layer. On the IBAD TiN layer on metal tape, we deposited approximately 80-nm-thick homo-epitaxial layer at about 600 °C to complete the conductive buffer stack. On the IBAD MgO layer on glass or metal tape, we first deposited a 40-nm-thick homo-epitaxial MgO layer at about 500 °C,[3,4] followed by a hetero-epitaxially deposited 150-nm-thick γ-Al$_2$O$_3$ layer at about 750 °C to complete the buffer layer stack. In both cases, the final layer of the buffer stack acted not only as an effective diffusion barrier, but also as a robust template layer for the hetero-epitaxial Si film growth at ~750 °C under an ambient chamber pressure of ~10^{-6} Torr. The Si growth rates varied between 0.5 to 2 nm/s. The p-type doping in the Si films was achieved by adding appropriate amounts of B to the Si source material, whereas n-type doping was achieved by simultaneous evaporation of P$_2$ from a compound dissociation cell during Si growth. A represantative architecture of two types of ACSi samples is shown in Figure 2.

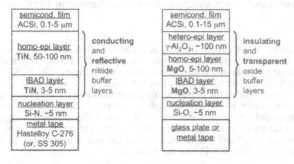

Figure 2. A cross-sectional schematic of ACSi film on metal tape with conducting and reflective buffer, and ACSi film on glass plate (or, metal tape) with insulating and transparent buffer.

DISCUSSION

The crystallographic orientation, epitaxial relationship, and microstructure of the multilayered samples were analyzed by X-ray diffraction (XRD), transmission electron microscopy (TEM), energy dispersive spectroscopy (EDS), scanning electron microscopy (SEM), and selected area diffraction (SAD). For these characterizations, we prepared samples with varying Si layer thicknesses (from 0.1 μm to 15 μm) and varying doping profiles under similar growth conditions.

Figure 3 is a composite micrograph of an ACSi film on oxide buffer on glass that shows a TEM cross-section (Figure 3(a)), SAD images of two grains A and B with the strongest contrast (Figures 3(b) and (c)), and misalignment between grains A and B (Figure 3(d)). This sample with about 4-μm-thick ACSi film had a total mosaic spread TMS (i.e., sum of in-plane and out-of-plane mosaic spreads) of ~3.5° (for definition of TMS, refer to refs. 4 and 7), measured with x-ray rocking curve and phi scans over an area of ~mm^2. Figure 3 indicates that although ACSi grains could have a high concentration of misfit dislocations and stacking faults, the grains are well aligned both in-plane and out-of-plane.

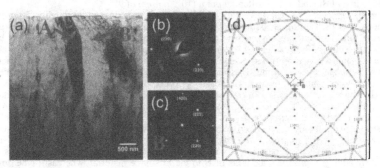

Figure 3. (a) Cross-sectional TEM of an ACSi film on glass substrate with transparent and insulating buffer; selected area diffraction of grain (b) A and (c) B showing single domain crystallinity along <001> with respect to substrate, and (d) grain misalignment of 3.7° along <110> direction.

Under optimized growth conditions, ACSi films demonstrate out-of-plane and in-plane mosaic spreads as low as 0.8° and 1.3°, respectively. Figure 4 compares TEM cross-section of a standard commercial silicon-on-sapphire (SOS) sample with two ACSi samples. The SOS sample (Figure 4(a)) and ACSi sample with low TMS of ~2° on oxide buffer on "smooth" Hastelloy C276 metal tape (Figure 4(b)) show similar misfit dislocations and stacking faults due to lattice mismatch between the Si film and the growth surface. In both cases, the defect density decreases (i.e., defects are annealed out) away from the interface. ACSi film depicted in Figure 4(c) on the other hand is representative of an ACSi film grown on a "rough" metal tape, in which case the rough surface of the substrate leads to additional defects that extend further into the Si film. This sample also shows larger TMS of ~12° for the ACSi film.

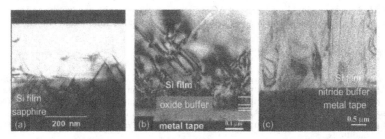

Figure 4. Cross-sectional TEM of (a) standard silicon-on-sapphire, (b) ACSi film on "smooth" metal tape with oxide buffer, and (c) ACSi film on "rough" metal tape with nitride buffer.

EDS line scans, shown in Figure 5, indicate that both the oxide and nitride buffer layers serve as good diffusion barriers with ~100 nm or less thickness. We believe that the uniform Fe intensity signal shown on the Si film side in Figure 5(c) is an artifact because of the strong background signal for Fe coming from the thick stainless steel substrate.

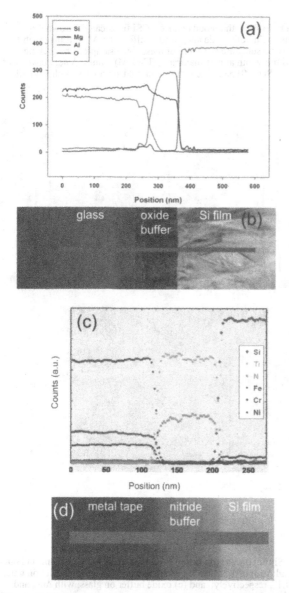

Figure 5. EDS line scan on an (a) ACSi film on oxide buffer on glass and (c) ACSi film on stainless steel tape with nitride buffer, with corresponding TEM images (b) and (d).

The XRD θ-2θ scans (Figure 6) show that good quality ACSi films can be grown epitaxially on metal tape and glass with both oxide and nitride buffer. The ACSi films exhibit (001) orientation perpendicular to the substrate surface. For these three samples, the rocking curves on the Si (004) peak yield full width at half-maximum (FWHM) values, $\Delta\omega_{Si}$, between 1° and 2°, whereas the φ-scans on the Si (220) peak show pure four-fold symmetry, with FWHM values, $\Delta\phi_{Si}$, between 2° and 4°.

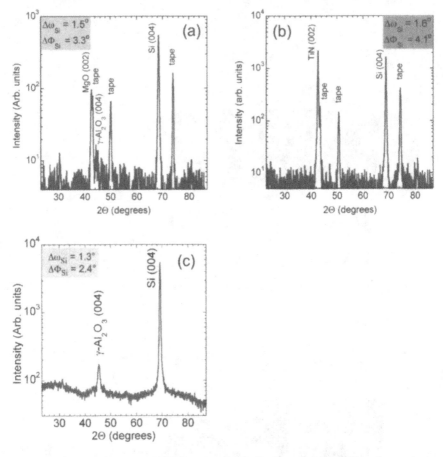

Figure 6. XRD θ-2θ scan of ACSi film on: (a) oxide buffer on tape, with ACSi in-plane and out-of-plane mosaic spreads ($\Delta\omega_{Si}$ and $\Delta\phi_{Si}$) of 1.5° and 3.3°, respectively; (b) nitride buffer on tape, with $\Delta\omega_{Si}$ and $\Delta\phi_{Si}$ of 1.6° and 4.1°, respectively; and (c) oxide buffer on glass, with $\Delta\omega_{Si}$ and $\Delta\phi_{Si}$ of 1.3° and 2.4°, respectively.

To study the effects of inter-grain alignment on the majority carrier transport in ACSi films on glass and metal tape, we prepared a series of samples with varying total mosaic spreads and patterned them for Hall mobility measurements. Figure 7 shows normalized Hall mobility vs total mosaic spread TMS for 0.2-to-4-μm-thick ACSi films. To be able to compare the carrier mobility of samples with different doping concentrations, we have normalized the carrier mobility of the films with respect to bulk single-crystal Si, and used a Hall factor of 0.8 to convert drift mobilities to Hall mobilities.[16,17] These results follow a common trend for ACSi films;[7,8,9] namely, for a given film thickness and doping concentration, as the TMS is reduced (i.e., the inter-grain alignment is improved) the carrier mobility increases. For example, a 0.4-μm-thick ACSi film on glass with a TMS of 4.2° showed Hall mobility of 47 cm^2/V.s (or, normalized mobility of 0.60) for a p-type doping concentration of 1.9×10^{18} cm^{-3}.

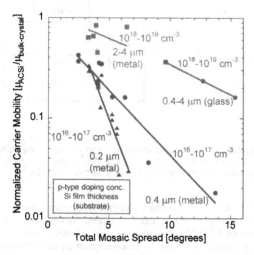

Figure 7. Normalized carrier mobility vs total mosaic spread *TMS* for ACSi films on glass and metal substrates (lines are guides to the eye).

Figure 8 shows the effect of temperature as well as of hydrogen passivation on the carrier mobility of ACSi films with different TMS. The details of this study have been published elsewhere.[9] As a reference, we have also included results on a similarly processed SOS sample. We observe that ACSi films with low TMS values show increasing mobility with decreasing temperature, similar to what is observed in single-crystalline Si films on single-crystal substrates (for example, SOS). This behavior is indicative of carrier scattering that is dominated by processes within the grains (see model depicted in Figure 8(b)). On the other hand, ACSi films with large-angle grain boundaries (i.e., with large TMS) show behavior that indicates carrier scattering that is dominated by grain boundary processes. In addition, ACSi films with low TMS and the SOS reference sample do not show any change in behavior with a hydrogen passivation treatment, whereas ACSi films with large TMS show appreciable improvement. This again supports our earlier results that ACSi films with predominantly low-angle grain boundaries have a low density of defects, and they don't need grain boundary modification (i.e., hydrogen passivation and/or grain recrystallization) to exhibit good majority carrier transport properties. In fact, a detailed study we have performed show that as the TMS decreases from ~15° to ~3°, the energy barrier height associated with the grain boundary scattering decreases monotonically from ~150 meV to less than 1 meV.[9,10]

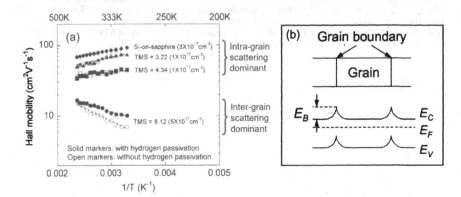

Figure 8. (a) Hall mobility vs temperature for ACSi films on oxide buffer on metal tape with varying total mosaic spread (TMS) and a Si-on-sapphire sample (lines are guides to the eye), (b) model used for the effect of grain boundaries, with associated energy barrier height E_B.[9]

We have also studied the effect of grain alignment on the surface electronic properties of ACSi films using metal-oxide-semiconductor (MOS) capacitor structures. For this work, we have fabricated MOS capacitors based on ACSi films with thermally-grown oxides on them. The details of the fabrication of and the capacitance-voltage (C-V) measurement on these structures are given elsewhere.[8] In Figure 9(a), C-V results show well defined regions of carrier accumulation, depletion, and inversion for a 4-μm-thick ACSi film with a light (p) to heavy (p+) doping profile from ACSi surface to substrate side. An accumulation capacitance appears at negative voltages indicating that the top layer of the ACSi film does not have a high density of defects. It also shows typical high- and low- frequency effects at 100 kHz and 1 kHz, respectively. The low frequency effect at 1 kHz under positive voltage bias is indicative of carrier inversion from p-type to n-type at the surface of the ACSi film. This type of inversion effect is crucial for operation of thin-film transistor (TFT) structures such as metal-oxide-semiconductor field-effect-transistors (MOSFETs). By repeating these measurements on other ACSi samples with varying TMS values and combining these results with a model based on the high-frequency method,[8] we determine the interface trap density at mid-gap (D_{it}) for ACSi films. The results in Figure 9(b) show that with improved alignment of Si grains (i.e., with decreasing TMS), ACSi films exhibit lower defect densities, asymptotically approaching levels observed in single-crystalline bulk Si and near-single-crystalline film Si (i.e., SOS).

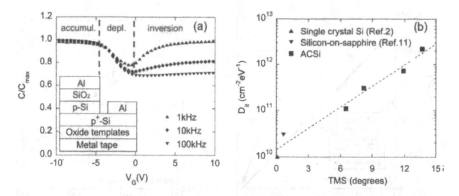

Figure 9. (a) Normalized capacitance vs gate voltage, showing carrier accumulation, depletion, and inversion in the Si film. Also shown is a schematic cross-section of the metal-oxide-semiconductor capacitor structure. (b) Interface trap density vs Total Mosaic Spread, showing decreasing trap density with improved grain alignment (dashed line is a guide to the eye).[8]

To examine electronic properties further, we prepared a vertical diode of the type n+/p/p+ where n+, p, and p+ refer to a heavily-doped n-type ($\sim10^{19}$ cm^{-3}), a lightly-doped p-type ($\sim5\times10^{16}$ cm^{-3}), and a heavily-doped p-type ($\sim10^{18}$ cm^{-3}) ACSi layer, respectively.[11] We used reactive ion etching to pattern ~0.5 mm x 0.5 mm mesas of n+ and p layers. We then evaporated ~200 nm of Ni and Al to form ohmic contacts on the n+ and p+ layers, respectively. The ACSi film for this diode had TMS $\sim 5.9°$.

Figure 10 shows the capacitance-voltage (C-V) characteristics of the diode prototype. A linear fit to the C^{-2} vs V curve yields a uniform p-layer doping of $\sim5\times10^{16}$ cm^{-3} and a built-in voltage V_{bi} of ~1.0 V.[11] This built-in voltage is similar to what is expected in a good crystalline-film Si diode with the above doping levels.[17] Normalized capacitance (C/C_0) and conductance (G/G_0) vs frequency of diodes can be used to estimate minority carrier lifetime in the lightly doped layer of a diode, as shown in Figure 10(b).[17] Combining this effective lifetime with an approximate estimate of electron mobility of ~150 cm^2/V.s in the p-layer (using the normalized hole mobility plot of Figure 7 and assuming that a similar dependence will hold for electrons) we estimate the effective diffusion length in the p layer of the diode prototype to be ~30 µm.

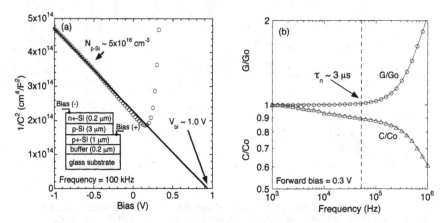

Figure 10. (a) Capacitance^{-2} vs bias (also included is a schematic cross-section of the 4.2-µm-thick ACSi film diode of the type n+/p/p+), and (b) normalized capacitance (C/Co) and conductance (G/Go) vs frequency for the diode, indicating minority carrier lifetime of ~3 µs in the p-layer.[11]

Figure 11 shows the photo-current response of the ACSi diode after deposition of an 80-nm-thick MgO layer that serves as a rudimentary anti-reflective (AR) coating. The schematic cross-section of the diode and the AR layer are illustrated in Figure 11(a). Figure 11(b) compares the external quantum efficiency (EQE) of the ACSi diode with that of a calibrated single-crystalline Si detector. This ACSi diode exhibits a strong response at ~450nm, with relatively sharp fall off above and below that wavelength. The short wavelength response is especially sensitive to surface effects and the n+/p junction properties, whereas the high-wavelength response is additionally influenced by the electronic properties of the back surface of the p+ layer and the photon absorption across the whole ACSi film. Crystalline Si, being an indirect bandgap semiconductor, requires ~50 µm of effective thickness to absorb long wavelengths (~1 µm). The fact that only 3-µm-thick p-Si absorber layer in this ACSi diode could lead to ~80% EQE at ~450 nm is encouraging because it should be possible to improve the higher wavelength response by using a thicker p-Si layer and employing an effective light-trapping scheme.

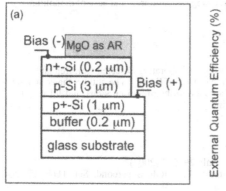 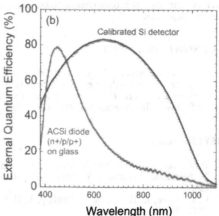

Figure 11. (a) Schematic cross-section of the ACSi diode of Fig. 10 after deposition of an MgO film as a rudimentary anti-reflector (AR), (b) external quantum efficiency vs wavelength for the ACSi diode with AR and a reference Si detector.

CONCLUSIONS

We have used an IBAD texturing technique to grow biaxially-oriented buffer layers on glass and metal substrates. The IBAD layer, which can be an oxide or nitride, constitutes a transformation of the amorphous/polycrystalline surface of the glass/metal to a highly ordered surface amenable to epitaxial growth of subsequent layers. We have built oxide-based and nitride-based epitaxial buffer stacks on these IBAD layers that allowed growth of ACSi films with varying concentrations of n- and p-type doping, and with thicknesses up to the explored range of 15 μm. Hall measurements confirmed that with improving crystallinity (i.e., with improving connectivity and alignment of the grains) in the ACSi film, the majority carrier mobility improves, asymptotically approaching values expected of single-crystalline Si films. We have also studied the minority carrier lifetime in ACSi films by fabricating an n+/p/p+ type diode. This diode prototype has provided evidence that our approach allows building of abrupt doping profiles of both p- and n-type crystalline Si films in situ on such non-single-crystalline substrates. Also, we have observed relatively long effective minority carrier lifetimes (~3 μs) and estimated diffusion lengths (~30 μm) in the 3-μm-thick p layer of such a diode. The same diode with the addition of an AR layer yielded EQE of ~80% at 450 nm. We note that this diode was based on an ACSi film with a non-optimal TMS of ~5.9°.

Most photovoltaics and electronics applications of semiconductor films on inexpensive non-single-crystalline substrates would benefit from improvements in performance and/or reductions in cost. Current technologies either use inexpensive but poor-performing amorphous semiconductor films, such as amorphous silicon,[18] or attempt to improve performance by transforming amorphous films to polycrystalline films by post-processing, such as solid phase crystallization,[19,20] or pulsed-laser crystallization.[21,22] These techniques lead to polycrystalline texture in the film with large-angle grain boundaries that could be detrimental to electronic and optical performance. Our approach uses an in situ process to grow grain-aligned Si films on glass and metal substrates, thus retains the advantages of conventional epitaxial growth techniques, such as better doping profile control, lower processing temperatures, and ease of monolithic device integration.

ACKNOWLEDGEMENTS

This work was funded by Los Alamos National Laboratory Directed Research and Development Project under the United States Department of Energy. The authors thank Woong Choi for his prior work, Brian Crone and Ian Campbell for helpful discussions, and Jeffrey O. Willis for his useful comments on the manuscript.

REFERENCES

1. P. N. Arendt, S. R. Foltyn, Mater. Res. Soc. Bull. **29**, 543 (2004).
2. Y. Iijima, K. Kakimoto, Y. Sutoh, S. Ajimura, T. Saitoh, Supercond. Sci. Tech. **17**, 264 (2004).
3. C. P. Wang, K. B. Do, M. R. Beasley, T. H. Geballe, R. H. Hammond, Appl. Phys. Lett. **71**, 2955 (1997).
4. A. T. Findikoglu, S. Kreiskott, P. M. te Riele, and V. Matias, J. Mater. Res. **19**, 501 (2004).
5. L. Dong, D. J. Srolovitz, G. S. Was, Q. Zhao, and A. D. Rollett, J. Mater. Res. **16**, 210 (2001).
6. L. Dong, L. A. Zepeta-Ruiz, D. J. Srolovitz, J. Appl. Phys. **89**, 4105 (2001).
7. A. T. Findikoglu, W. Choi, V. Matias, T. G. Holesinger, Q. X. Jia, D. E. Peterson, Adv. Mater. **17**, 1527 (2005).
8. W. Choi, J. K. Lee, A. T. Findikoglu, Appl. Phys. Lett. **89**, 262111 (2006).
9. W. Choi, A. T. Findikoglu, M. J. Romero, M. Al-Jassim, JMR **22**, 821 (2007).
10. A. T. Findikoglu, W. Choi, M. Hawley, M. J. Romero, K. M. Jones, M. M. Al-Jassim, in *Progress in Advanced Materials Research* (Ed: N. H. Voler, Nova Science Publishers, Hauppauge, New York, 2007), Ch. **6**.
11. A. T. Findikoglu, O. Ugurlu, and T. G. holesinger, Mater. Res. Soc. Symp. Proc. 1066 (in press).
12. K. Yamamoto, IEEE Trans. Electron Dev. **46**, 2041 (1999).
13. A. G. Aberle, P. I. Widenborg, D. Song, A. Straub, M. L. Terry, T. Walsh, A. Sproul, P. Campbell, D. Inns, B. Beilby, M. Griffin, J. Weber, Y. Huang, O. Kunz, R. Gebs, F. Martin-Brune, V. Barroux, S. H. Wenham, "Recent Advances in Polycrystalline Silicon Thin-Film Solar Cells on Glass at UNSW", presented at *Thirty-First IEEE Photovoltaic Specialists Conference* (Lake Buena Vista, FL, USA, 2005).
14. J. H. Werner, R. Dassow, T. J. Rinke, J. R. Kohler, R. B. Bergmann, Thin Sol. Films **383**, 95 (2001).
15. C. W. Teplin, D. S. Ginley, H. M. Branz, J. Non-Cryst. Solids **352**, 984 (2006).
16. J. F. Lin, S. S. Li, L. C. Linares, K. W. Teng, Solid State Electron. **24**, 827 (1981).
17. S. M. Sze, *Physics of Semiconductor Devices* (Wiley-Interscience, New York, 1981).
18. S. Toshiharu, J. Appl. Phys. **99**, 11 (2006).
19. L. Haji, P. Joubert, J. Stoemenos, N. A. Economou, J. Appl. Phys. **75**, 3944 (1994).
20. R. B. Bergmann, J. Kohler, R. Dassow, C. Zaczek, J. H. Werner, Physica Status Solidi **A166**, 587 (1998).
21. J. S. Im, R. S. Sposili, M. A. Crowder, Appl. Phys. Lett. **70**, 3434 (1997).
22. M. Tai, M. Hatano, S. Yamaguchi, T. Noda, S. K. Park, T. Shiba, M. Ohkura, IEEE Trans. Electron. Devices **51**, 934 (2004).

Mater. Res. Soc. Symp. Proc. Vol. 1150 © 2009 Materials Research Society   1150-RR04-01

Development of conducting buffer architectures using cube textured IBAD-TiN layers

Ruben Hühne[1], Konrad Güth[1], Martin Kidszun[1], Rainer Kaltofen[1], Vladimir Matias[2], John Rowley[2], Ludwig Schultz[1] and Bernhard Holzapfel[1]

[1]IFW Dresden, P.O. Box 270116, D-01171 Dresden, Germany

[2]Superconductivity Technology Center, Los Alamos National Laboratory, Los Alamos, NM 87545, USA

ABSTRACT

Ion-beam assisted deposition (IBAD) offers the possibility to prepare thin textured films on amorphous or non-textured substrates. In particular, the textured nucleation of TiN is promising for the development of a conducting buffer layer architecture for YBCO coated conductors based on the IBAD approach. Accordingly, cube textured IBAD-TiN layers have been deposited reactively using pulsed laser deposition on Si/Si_3N_4 substrates as well as on polished Hastelloy tapes using different amorphous seed layers. Metallic buffer layers such as Au, Pt or Ir were grown epitaxially on top of the TiN layer showing texture values similar to the IBAD layer. Smooth layers were obtained using a double layer of Au/Pt or Au/Ir. Biaxially textured YBCO layers were achieved using $SrRuO_3$ or Nb-doped $SrTiO_3$ as a conductive oxide cap layer. Finally, different amorphous conducting seed layers were applied for the IBAD-TiN process. Highly textured TiN films were achieved on amorphous $Ta_{0.75}Ni_{0.25}$ layers showing a similar *in-plane* orientation of about 8° as on standard seed layers.

INTRODUCTION

Ion beam assisted deposition (IBAD) of biaxially textured buffer layers is one of the major routes for the preparation of suitable templates for YBCO coated conductors [1,2]. In particular, the deposition of highly textured MgO layers has gained a lot of interest as the desired cube texture is already created during nucleation within the first 10 nanometres making this process very time-efficient [3]. It was shown recently that other materials having a rocksalt structure, for example TiN, can be textured in a similar way [4]. One major advantage of TiN is its good electrical conductivity. This would enable the realization of a conductive buffer architecture within the IBAD approach leading to an electrical connection between the superconducting layer and the thick metal substrate in order to avoid thermal destruction of the superconductor in case of an overcurrent situation.

So far only non-conducting buffer architectures have been prepared based on textured IBAD-TiN layers [5]. Two major challenges need to be solved in order to realize a completely conducting buffer layer stack: (i) an electrically conductive amorphous or nanocrystalline seed layer has to be applied on the metal tape; (ii) additional conducting buffers have to be deposited on the IBAD layer in order to reduce the significant lattice mismatch between TiN (lattice parameter a = 0.424 nm) and YBCO (a ≈ b ≈ 0.386 nm) and to avoid the oxidation of the nitride layer. Different noble metals were tested for their suitability as buffers on TiN. Among them, gold (a = 0.408 nm) shows a good epitaxial growth on the nitride. Furthermore, Pt (a = 0.392 nm) and Ir (a = 0.386 nm) were used in order to reduce the lattice misfit. In addition, conducting oxide SrRuO₃ (a ≈ 0.393 nm) and Nb-doped SrTiO₃ layers (Nb:STO, a = 0.391nm) were used as cap layer prior to the deposition of the YBCO in a high oxygen pressure. Finally, different amorphous metallic seed layers were tested for the preparation of highly textured IBAD-TiN layers.

EXPERIMENT

Standard pulsed laser deposition (PLD) was used to prepare biaxially textured TiN films and subsequent buffer layers. These layers were grown in a high vacuum chamber with a background pressure of about 1×10^{-6} mbar. A Lambda Physik KrF excimer laser ($\lambda = 248$ nm) was applied on pure metals as well as on SrRuO₃ and Nb:SrTiO₃ with laser energy densities of 2-8 J/cm^2 and laser pulse repetition rate of 1-10 Hz. An rf plasma source provides the assisting ion beam with a mean ion beam energy of 800 eV using a gas mixture of argon and nitrogen, impinging the substrate surface under an angle of 45° to the substrate normal.

Commercially available Si/Si₃N₄ substrates as well as Hastelloy® C276 substrates coated with an amorphous Al₂O₃ seed layer were cut to pieces of 10 mm x 10 mm and introduced in the IBAD chamber. In the first step, a biaxially textured TiN thin film was grown using the IBAD process at a substrate temperature of about 350°C. The thickness of the TiN was increased using homoepitaxial growth at a substrate temperature of 700°C in order to preserve the cube texture. Afterwards, Au, Pt, Ir, SrRuO₃ or Nb:STO layers were prepared by PLD using a temperature of about 600°C. The texture evolution during film growth was observed *in situ* using a STAIB Instruments reflection high energy electron diffraction (RHEED) system, where an electron beam with an energy of 30 keV hits the substrate surface under a grazing incidence angle of 0.5-1.5°. The diffraction pattern was monitored by a CCD camera. Finally, a 300 nm thick YBCO layer was deposited on the grown buffer architecture using PLD in a separate high vacuum chamber at a background pressure of 30 Pa Oxygen and a substrate temperature of 810°C. More details to the YBCO deposition can be found elsewhere [6].

The texture was studied quantitatively by X-ray diffraction (XRD) in order to determine the full width at half maximum (FWHM) values of the *in-plane* mosaic spread of each layer. The surface structure and the local roughness were measured by scanning electron microscopy (SEM) and atomic force microscopy (AFM). Finally, resistive measurements on unpatterned samples were used to determine the critical temperature T_c of the superconducting layer.

RESULTS

Deposition of noble metals on IBAD TiN

For a first set of experiments commercially available Si-wafers covered with a smooth amorphous Si₃N₄ coating were used to develop a conducting buffer architecture based on IBAD-TiN. A sharp nucleation texture was observed within the first few nanometres of TiN film grown on this substrate using ion-beam assisted pulsed laser deposition [4]. The ion beam was switched off, when the diffraction pattern indicated a strong biaxial alignment in order to avoid the destruction of the desired cube texture during further ion beam assistance. Instead, the thickness of the cube textured TiN layer was increased using a standard reactive pulsed laser deposition at a temperature of 700°C in a nitrogen-containing atmosphere. The real-time RHEED pattern confirmed a homoepitaxial growth with sharper diffraction spots for an increasing film thickness. The rms-roughness of such homoepitaxial grown TiN layers are typically between 1 and 2 nm for film thicknesses below 50 nm.

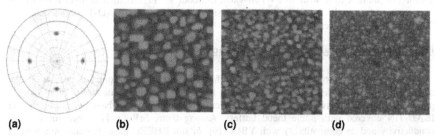

(a)   **(b)**   **(c)**   **(d)**

Figure 1. (a) X-ray (111) pole figure of a 100 nm thick Au layer on Si/Si$_3$N$_4$/IBAD-TiN/TiN; Surface morphology measured by AFM (2 μm x 2 μm) of: (b) ~100 nm Au; (c) ~20 nm Au and ~40 nm Pt; (d) ~20 nm Au and ~20 nm Ir on Si/Si$_3$N$_4$/IBAD-TiN/TiN, respectively.

Different noble metals were deposited on the highly textured TiN layer. It was found during *in-situ* RHEED measurements that Au grows epitaxially on TiN. XRD measurements confirmed the strong cube texture with an *in-plane* FWHM of 9.4° using the (111) planes (Fig. 1(a)). The results of AFM measurements on the gold layer revealed a high roughness with an rms-value of about 21 nm for about 100 nm thick films (Fig. 1(b)). Therefore, the thickness of the Au layer needs to be restricted. But nevertheless, the Au layer was successfully used to adopt the lattice misfit for the following buffer layers as the direct deposition of other noble metals as Pt or Ir on TiN led only to a disturbed epitaxial growth. As a result, double layers of Au(~20 nm)/Pt(~40 nm) or Au(~20 nm)/Ir(~20 nm) show an improved rms-roughness of 6.5 nm and 4.5 nm, respectively (Fig. 1(c)+(d)).

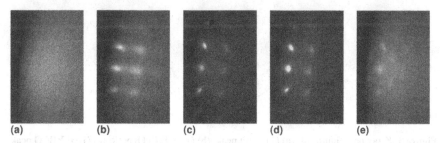

(a)   (b)   (c)   (d)   (e)

Figure 2. *In situ* RHEED pattern showing the surface texture of a complete buffer architecture deposited on a Hastelloy/a-Al$_2$O$_3$ substrates: (a) amorphous structure of the Al$_2$O$_3$ layer at 350°C; (b) TiN after homoepitaxial growth; (c) Au; (d) Pt; (e) SrRuO$_3$.

The developed buffer architecture was transferred in the next step to technical substrates. Electropolished Hastelloy tapes covered with an amorphous Al$_2$O$_3$ layer were used for this purpose [7]. In-situ RHEED measurement revealed the highly textured growth of the IBAD TiN as well as the epitaxial transfer of the texture to the Au and Pt layer (Fig. 2 (a)-(d)). The rms-roughness of the Pt layer was with about 4 nm slightly better compared to the results on the Si/Si_3N_4 substrate. XRD measurements on the Au layer of the grown buffer architecture revealed a strong *in-plane* texture with an FWHM-value of about 9° (Figure 3(a)). Similar results have been obtained recently on Hastelloy/a-Y_2O_3 substrates using an IBAD-TiN/homoepitaxial-TiN/Au/Ir buffer architecture [8].

Preparation of electrically conductive oxide cap layers

In general, an oxide cap layer is preferred to enable the epitaxial growth of the superconductor as the final YBCO layer needs to be deposited at about 800°C in an oxygen pressure of 0.3 mbar. Therefore, different electrically conductive oxides have been tested on the IBAD-TiN covered with noble metal buffers. Among them, $SrRuO_3$ is known for its high conductivity and its compatibility with YBCO [9]. *In situ* RHEED measurements indicated an epitaxial growth of the oxide buffer layer on Pt (Fig. 2(e)). However, the broad spots as well as the high background signal indicate that further optimization of the deposition conditions are necessary to achieve better textures. This is confirmed by XRD pole figure measurements, where an *in-plane* FWHM of about 20° was found for this buffer layer.

Nevertheless, the buffer architecture was used for the deposition of a final YBCO layer. The results of the XRD measurement on the superconducting layer shown in Fig. 3(c) reveal a broad *out-of-plane* and *in-plane* distribution as well as an additional fibre texture. Furthermore, no superconducting transition was observed above 77 K for the grown film. In contrast, YBCO layers grown on IBAD-TiN based conductive buffer architectures showed a resistively measured superconducting transition of about 88 K, if Nb:STO is used as cap layer [8].

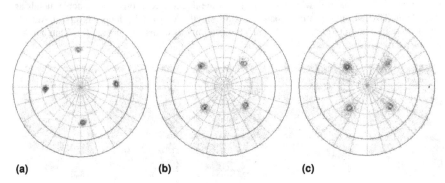

(a)   (b)   (c)

Figure 3. X-ray pole figures of: (a) (111) Au peak; (b) (110) $SrRuO_3$ peak; (c) (103) YBCO peak in a Hastelloy/Al_2O_3/IBAD-TiN/TiN/Au/Pt/$SrRuO_3$/YBCO buffer architecture.

Application of conducting seed layers for IBAD-TiN

A major prerequisite for the texturing of materials with a rocksalt structure using IBAD is an amorphous or nanocrystalline seed layer. Such a layer must also be electrically conductive, if completely conductive buffer architecture is desired. Therefore, different materials were tested to see, if they were suitable as conductive seed layers for the IBAD-TiN process. At first, 50 nm thick amorphous CoFeB layers were deposited using sputtering at room temperature on Si/Si_3N_4 substrates. The surface of these layers remained in an amorphous state as indicated by the diffuse RHEED pattern in Fig. 4(a). The typical nucleation cube texture was observed for IBAD-TiN layers deposited at 400°C, which could be preserved to higher thicknesses (Fig. 4(b)). However, the circular elongation of the diffraction spots indicates that the *in-plane* alignment is qualitatively not very god. This is confirmed by the RHEED pattern and the XRD pole figure of the gold buffer layer, where an additional fibre texture is observed (Fig. 4(c)+(d)).

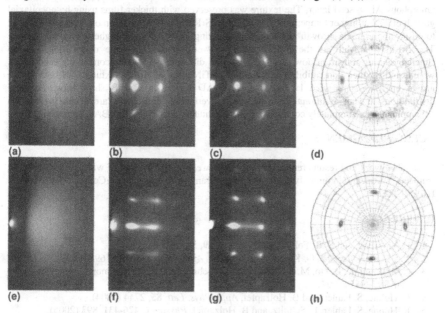

Figure 4. *In situ* RHEED pattern showing the surface texture of: (a) CoFeB on Si/Si_3N_4 at 400°C; (b) TiN after the IBAD step and the homoepitaxial growth; (c) gold deposited on top of this structure; (d) corresponding Au (111) pole figure; (e)-(h) shows the same layer sequence for a $Ta_{0.75}Ni_{0.25}$ seed layer deposited on Si/Si_3N_4 with an IBAD growth temperature of 250°C.

Amorphous $Ta_{0.75}Ni_{0.25}$ seed layers with a thickness of 50 nm were prepared in a second set of experiments using co-sputtering from elemental targets at room temperature. Such amorphous metallic layers were investigated recently as barrier layers for Cu interconnections in Si-based microelectronics [10]. The deposited layers showed an amorphous surface structure at elevated temperature as shown in the RHEED pattern in Fig. 4(e). Highly textured IBAD-TiN layers were achieved on this seed layer as already described in detail in a previous publication [11]. This texture was preserved using homoepitaxial growth leading to *in-plane* FWHM values of about 8° for the gold buffer layer (Fig. 4(f)-(h)). The results indicate that amorphous TaNi seed layers are suitable templates for the ion-beam assisted pulsed laser deposition of TiN.

CONCLUSIONS

In summary, a conductive buffer architecture for coated conductors was developed on the basis of IBAD-TiN. Noble metals were used as first buffer layers. A double layer of Au/Ir or Au/Pt showed undisturbed epitaxial growth and a low surface roughness. The IBAD-TiN process was successfully transferred afterwards to technical substrates using Hastelloy with an amorphous Al_2O_3 seed layer. The texture was preserved with thicker films using homoepitaxial growth of TiN at higher temperatures. In addition, $SrRuO_3$ was tested as an oxide cap layer on Pt to ensure the epitaxial growth of the superconducting YBCO layer at higher oxygen pressures. The $SrRuO_3$ as well as the YBCO layer showed a broad in-plane distribution and no superconducting transition down to 77 K. Finally, different amorphous conducting seed layers were tested for their compatibility with the IBAD-TiN process. Highly textured TiN layers were grown on $Ta_{0.75}Ni_{0.25}$ seed layers, whereas IBAD layers on CoFeB show broad *in-plane* distributions and an additional fibre texture. Nevertheless, the results are promising for the development of a completely conductive buffer architecture based on the IBAD approach.

ACKNOWLEDGMENTS

The work and results reported in this letter were partially obtained with research funding from the European Community under the Sixth Framework Programme Contract No. 516858: HIPERCHEM.

REFERENCES

1. P. N. Arendt and S. R. Foltyn, *MRS Bulletin* **29**, 543 (2004).
2. Y. Iijima, N. Tanabe, O. Kohno, and Y. Ikeno, *Appl. Phys. Lett.* **60**, 769 (1992).
3. C. P. Wang, K. B. Do, M. R. Beasley, T. H. Geballe, and R. H. Hammond, *Appl. Phys. Lett.* **71**, 2955 (1997).
4. R. Hühne, S. Fähler, and B. Holzapfel, *Appl. Phys. Lett.* **85**, 2744 (2004).
5. R. Hühne, S. Fähler, L. Schultz, and B. Holzapfel, *Physica C* **426-431**, 893 (2005).
6. R. Hühne, V. S. Sarma, D. Okai, T. Thersleff, L. Schultz, and B. Holzapfel, *Supercond. Sci. Technol.* **20**, 709 (2007).
7. S. Kreiskott, P. N. Arendt, J. Y. Coulter, P. C. Dowden, S. R. Foltyn, B. J. Gibbons, V. Matias, and C. J. Sheehan, *Supercond. Sci. Technol.* **17**, S132-S134 (2004).
8. K. Güth, R. Hühne, V. Matias, J. Rowley, T. Thersleff, L. Schultz, and B. Holzapfel, *IEEE Trans. Appl. Supercond.* **19**, accepted (2009).
9. R. Hühne, D. Selbmann, J. Eickemeyer, J. Hänisch, and B. Holzapfel, *Supercond. Sci. Technol.* **19**, 169 (2006).
10. J. S. Fang, T. P. Hsu, and H. C. Chen, *J. Electr. Mater.* **36**, 614 (2007).
11. R. Hühne, K. Güth, M. Kidszun, R. Kaltofen, L. Schultz, and B. Holzapfel, *J. Phys. D: Appl. Phys.* **41** (2008) 245404.

Mater. Res. Soc. Symp. Proc. Vol. 1150 © 2009 Materials Research Society   1150-RR04-03

Texture Evolution in Ion-Beam-Assist Deposited MgO

Vladimir Matias, Marcel Hoek, Jens Hänisch, and Alp T. Findikoglu
Superconductivity Technology Center, Mail Stop T004, LANL, Los Alamos, NM 87545, U.S.A.

ABSTRACT

We examine crystalline-texture evolution during ion-beam assisted deposition (IBAD) of MgO thin films. Flexible metal tapes are used as substrates in a reel-to-reel tape deposition system. Continuously moving substrates allow for a linear combinatorial experimental method. IBAD growth is monitored *in situ* in real-time with reflection high energy electron diffraction. We have demonstrated that in-plane crystalline texture evolution in IBAD-MgO scales with deposition rate. At high ion currents an in-plane texture full-width half maximum (FWHM) of 10° can be achieved in less than 1 second, and 6° in 2.2 seconds. An x-ray texture analysis within a beam reference frame for the rocking curves is presented. We suggest that this reference frame is the natural one for IBAD texture evaluation and, furthermore, that the FWHM for these rocking curves are more easily extracted than those within the film reference frame.

INTRODUCTION

Among the numerous materials that are amenable to ion-beam-assisted deposition (IBAD) crystalline grain alignment during film growth, particular scientific and application interests are in the ones that can form texture at nanometer thicknesses, i.e. very rapidly during deposition. Ion-beam texturing in MgO during early stages of film growth was indicated previously to start at grain nucleation.[1,2] One can identify three different texture development regions in MgO film growth under IBAD. The first stage, where biaxial texture first appears, is during grain nucleation. There is indication of a phase transition from an amorphous phase to a crystalline one in this region.[2] With additional IBAD texture continues to improve by grain alignment, but up to a certain point where it saturates and begins to degrade. Further improvement in crystalline alignment can be achieved by a third stage of epitaxial overgrowth, with either homoepitaxial or heteroepitaxial deposition at elevated temperature; see Fig. 1 for an illustration of these regions.

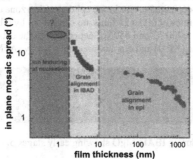

Figure 1. In-plane texture of MgO as a function of film thickness through the three different texture evolution regions, shown on logarithmic scales.

In this paper we review some of our recent results regarding a search for a better understanding of the various texture development regions and discuss texture analysis from x-ray diffraction data and how it might be improved.

EXPERIMENT

As our substrate we use electropolished Hastelloy C-276 Ni-alloy substrate, 100 μm thick,[3] with a typical surface roughness RMS value of about 0.5 – 1 nm on a 5 μm x 5 μm area. The substrate roughness is known to influence the texturing process,[4] and therefore we take great care to prepare smooth starting surfaces which we do in our laboratory by electropolishing. In the experiments we use a continuous feed of substrate, in lengths of 5 – 10 m, provided by reel-to-reel tape transport inside the vacuum deposition system. A linear combinatorial experimental approach is used for preparation of some of the samples, such as to obtain thickness profiles, or 'wedges', by moving tape into a deposition zone and then stopping deposition; see Ref. 5 for a more detailed description. The substrate is ion etched in the vacuum chamber for 50 s at 750 V and with an ion current density of 0.4 mA/cm^2. We use a Kaufmann 22-cm rf ion source with a neutralized beam of Ar ions incident at 45° to the sample normal for all the work described here. The substrate etching is used to eliminate any contaminant residue from the surface, as we have directly observed by *in situ* surface ion scattering.[5] Y_2O_3 is deposited by electron-beam evaporation at a rate of 0.2 nm/s for a total thickness of 14 nm. The evaporation rate is monitored and controlled by a quartz crystal microbalance (QCM). An ion etch is then applied for 25 seconds with identical parameters as stated above. This step etches the Y_2O_3 layer to a total thickness of less than 10 nm and disorders the surface as observed *in situ* by reflection high-energy electron diffraction (RHEED). The Y_2O_3 layer is used as a 'bed' layer for the IBAD-MgO texturing process.

IBAD-MgO is deposited by electron-beam sublimation from an MgO source and with an ion beam assist beam of energy 750 – 1500 V and ion-beam current density of 0.4 – 3 mA/cm^2. The sample is at ambient temperature and loosely lies on a water-cooled copper block. Temperature is estimated to be less than 100 °C in all cases. The IBAD process is monitored with two QCM's. One QCM is exposed to just the MgO flux (QCM 1) and the second one is exposed to both MgO and the ion flux (QCM 2). This is done to monitor the ratio of ions-to-molecules.[6] RHEED is also used to monitor the texture formation in real time. Figure 2 shows a typical texture evolution in IBAD as monitored by RHEED. A homoepitaxial layer of MgO of 120 – 1500 nm thickness is deposited *in situ* in the epi deposition zone of the IBAD deposition system.[5] MgO deposition rates of 0.1 to 25 nm/s are carried out by e-beam sublimation at sample temperatures between 400 °C and 650 °C, depending on the deposition rate. Samples are analyzed by x-ray diffraction (XRD) and atomic force microscopy (AFM).

Figure 2. RHEED images during IBAD-MgO showing early stages of texture evolution.

RESULTS

IBAD-MgO thickness wedges as in Ref. [5] were prepared under various IBAD growth conditions and identical homoepitaxial MgO layers were deposited on top for comparison. The samples were then characterized for their crystalline texture as determined by XRD measurements of the mosaic spreads. Data as a function of thickness is, in this case, easily obtained by measuring along the length of the tape sample. There are three independent angles for measuring the mosaic spreads, for which we chose two out-of-plane and one in-plane for the 'film' reference frame (FRF); see our discussion below. We measure the MgO (002) rocking curves in two directions to determine the full-width half maxima (FWHM) for the two out-of-plane angles, $\Delta\omega_1$ and $\Delta\omega_2$. For the data presented here we average the two values to obtain an average out-of-plane $\Delta\omega$ FWHM. The (202) pole figures are measured and the average $\Delta\phi$ FWHM from the four poles is deduced by taking out the contributions from the appropriate $\Delta\omega$'s, according to the formula: $\Delta\phi = (\Delta\phi_{202}^2 - \Delta\phi_\omega^2)^{1/2}$.[7] This formula assumes independent Gaussian diffraction peaks.[8] The XRD measurements and the FWHM determination are performed on the full MgO layer, i.e. IBAD and homoepitaxial layers together. Figure 3 shows the texture evolution results for different IBAD MgO deposition rates while varying the ion beam current concurrently to keep the ion-to-molecule ratio approximately constant. In this case the homoepitaxial layer was 120 nm thick. Across a wide range of IBAD deposition times and deposition rates the data can be made to scale if we plot the texture FWHM as a function of the ion-etched thickness on QCM2; see Ref. 9 for a more complete presentation of the results. We have developed an empirical quantification of the texture evolution curves in both IBAD and homoepitaxial layers.[9] The data can be reasonably fit to an exponential decay as shown in Fig. 3.

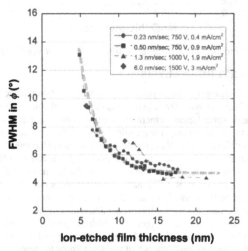

Figure 3. In-plane texture for the IBAD-Epi composite as a function of IBAD-MgO layer thickness for various IBAD deposition rates/ion currents. The data are scaled to the ion-etched film thickness based on QCM1-QCM2 readings.

We prepared a similar wedge sample for the homoepitaxial layer, i.e. in this case the IBAD layer was of constant thickness and homoepitaxial MgO thickness was varied. Figure 4 shows the texture evolution as a function of epi-layer thickness. We see a similar functional behavior and the data are fit again to the exponential decay. The texture continues to improve in the epi-layer. The best texture we attained thus far in the MgO layer on polished metal tape has an in-plane FWHM of 1.6° for a 1.5 μm thick film.

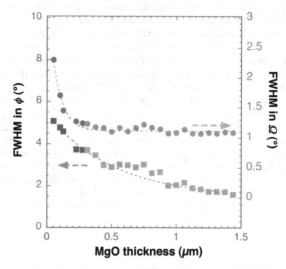

Figure 4. Texture as a function of homoepitaxial film thickness for Epi-MgO grown at 600°C.

DISCUSSION OF TEXTURE ANALYSIS

It is nontrivial to determine the in-plane mosaic misorienitation since true in-plane XRD measurements are difficult to make. Thus, in order to determine the in-plane mosaic spreads one typically makes a mosaic spread measurement on an off-axis peak and then subtracts out the out-of-plane contribution to derive the in-plane mosaic spread.[7] Figure 5 shows a typical {202} XRD pole figure for an IBAD-MgO (plus epi-MgO) sample. It is characterized by asymmetric peaks that are not equivalent in shape or width. The peak in the direction of the ion beam is typically narrower than the others, and the one away from the beam is the broadest. The derived in-plane misorientations, following Ref. [7], have different values outside of measurement error, even though they should be the same, indicating that the independent Gaussian peak assumption is not correct. One can average the four values to get a 'mean' in-plane value, and this is often done experimentally. However the data leaves some ambiguity since the four angles are not the same without a good physical explanation, and thus the derivation of the in-plane misorientation is questionable. The fundamental problem is that we don't know the true rocking curve peakshapes and how the different rocking curve components couple for interim angles. This implies that the derived in-plane misorientation values are not accurate, due to the questionable analysis that is often applied. We suggest that "in-plane" symmetry is not natural for the IBAD texture and that there is a better way to do this analysis that is not ambiguous.

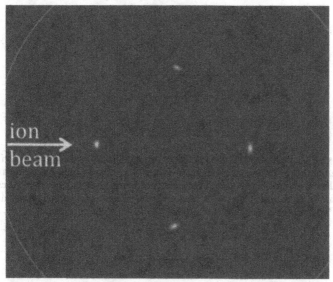

Figure 5. {202} XRD pole figure of the IBAD-MgO sample as measured by XRD.

Our approach is to use a different reference frame for the analysis of the mosaic misorientations. Instead of using the 'film' reference frame (FRF), i.e. the one in-plane $(\Delta\phi)$ and two out-of-plane angles $(\Delta\Omega_1$ and $\Delta\Omega_2)$ to characterize mosaic spreads of the sample, we propose to use the reference frame that utilizes the ion beam direction, in this case at 45° to the film normal. We call this the 'beam' reference frame (BRF). We take the beam direction as one of the BRF principal axes, $[202]_{//}$. The second principal axis is easily taken along the <202> direction orthogonal to the beam in reciprocal space and pointing 180° azimuthally out of the film plane; see Fig. 6, $[202]_{\perp}$. The misorientations are measured and indicated with the rocking angles that twist around these axes, in this case $\Delta\Omega_{202//}$ and $\Delta\Omega_{202\perp}$. The third orthogonal axis is then in the film plane, orthogonal to the other two; the twist around that axis is shown in Figure 6 as $\Delta\Omega_1$. Using the BRF simplifies data collection and analysis since all of these three orthogonal rocking curves can be measured **directly** and without any need for deconvolution of the components. Furthermore we believe that this data is inherently more accurate and physical for this system. Note in Fig. 5 that the {202} poles perpendicular to the ion beam plane are curved out of the film normal, but instead centered around the ion beam direction. Also, as noted earlier, the in-beam spread is much smaller than the perpendicular one. This implies that these two directions are fundamentally different for the IBAD process and we believe should be analyzed separately.

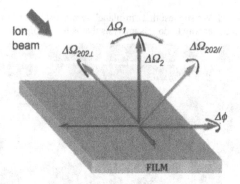

Figure 6. Reference frames as described in the text. In the film reference frame, shown in blue, one uses the film out-of-plane axis as one of the principal axes, and two in-plane film axes defined by the off-axis ion beam. In the beam reference frame, shown in green, one uses the beam axis as one of the principal axes, the opposing <202> axis as the second principal axis, and the in-plane axis that is perpendicular to the other two for the third principal axis.

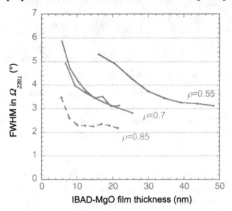

Figure 7. Rocking curves for the (202) peak in the beam direction plotted for a variety of ion-to-molecule ratios, represented by the different ρ values. The two red lines represent two different sets of data for the same ratio.

Figure 7 shows some preliminary IBAD FWHM texture data for the rocking curves of the {202} peak in the beam direction, as a function of thickness, for a number of ion-to-molecule ratios. The ρ values reflect the ratio of quartz crystal monitors, as discussed in Ref. 6, which are proportional to the ion-to-molecule ratios. The texture under IBAD improves, but with a different slope for different ion-to-molecule ratios. While the trend in the data is the same as that from the deduced $\Delta\phi$ values, we believe that these {202} rocking curve data are more accurate and will better reflect the nature of the IBAD texturing phenomenon.

CONCLUSIONS

Three different regions for biaxial texture formation and development are clearly identified in IBAD of MgO. In the first region, or the nucleation layer, there is as yet no texture developed, until at one point when it emerges. Beyond that, texture improves during IBAD up to a saturation value. The texture can improve further, albeit more slowly, by epitaxial overgrowth at higher temperatures. We achieved a best in-plane FWHM misorientation value of 1.6°. We propose a 'beam reference frame' for analysis of mosaic spreads which uses the beam direction as one of its principal axes. This method simplifies the measurement of the mosaic spreads and might give more insight into the physics of the process.

ACKNOWLEDGMENTS

The authors acknowledge many colleagues who influenced the development of IBAD-MgO texturing and numerous discussions with them, especially Robert Hammond (Stanford U), Paul Arendt, Randy Groves, Liliana Stan (LANL), Ruben Hühne (IFW Dresden), and Judy Wu (U Kansas). This work is supported by the Department of Energy, Office of Electricity Distribution & Energy Reliability.

REFERENCES

1. C.P. Wang, K.B. Do, M.R. Beasley, T.H. Geballe, and R.H. Hammond, Appl. Phys. Lett. 71, 2955 (1997).
2. Rhett T. Brewer and Harry A. Atwater, Appl. Phys. Lett. 80, 3388 (2002).
3. Sascha Kreiskott, Paul N. Arendt, Lawrence E. Bronisz, Steve R. Foltyn, and Vladimir Matias, Supercond. Sci. & Technol. 16, 613 (2003).
4. V. Matias, J. Hänisch, E.J. Rowley, C. Sheehan, P.G. Clem, N. Kumasaka, and I. Kodaka, Mater. Res. Soc. Symp. Proc. Volume 1001E, Warrendale, PA, 2007, No. 1001-M04-02.
5. Vladimir Matias and Brady J. Gibbons, 78, 072206 (2007).
6. Alp T. Findikoglu, Sascha Kreiskott, Paul M. te Riele, and Vladimir Matias, J. Materials Research 19, 501 (2004).
7. E.D. Specht, A. Goyal, D.F. Lee, F.A. List, D.M. Kroeger, M. Paranthaman, R.K. Williams, and D.K. Christen, Supercond. Sci. Technol. 11, 945 (1998).
8. V. Srikant, J.S. Specht, and D.R. Clarke, J. Appl. Phys. 82, 4286 (1997).
9. V. Matias, J. Hänisch, E.J. Rowley, and K. Güth, J. Materials Research, 24, 125 (2009).

Round Table Discussion: Texturing Mechanisms in Thin Films
Symposium RR, 4:15 pm December 2, 2008

Panel: James Harper (Univ of New Hampshire), Robert Hammond (Stanford Univ), David Srolovitz (Yeshiva Univ) and Harry Atwater (Caltech)
Session chair: Vladimir Matias (LANL)

Possible discussion topics listed by the session chair:
- Understanding of IBAD texturing mechanisms
  - Texturing at film nucleation (as in MgO, TiN)
  - Texturing during/after/alternating with film growth
- Understanding its applicability to various materials
- Other methods of texturing such as inclined deposition
- Ultimate limits on biaxial grain alignment
- Areas for further research
- Applications of textured templates

James Harper:
This is an open discussion and I would like to go directly to this topic: What do we think are the ultimate limits on biaxial alignment in the ion beam systems we have right now? There is a certain divergence in the beam; do people feel that this is setting the limit or are we talking about a material property in the film itself that is setting the limit?

Robert Hammond:
I think it is Srolovitz's fault: his simulations showed that you cannot get better than ±15° (30°) based on ion channeling itself, so how do we get beyond that? The problem is how do you get grains to align further?
My question to Harry is, how good do you need it to be for your (plasmonic) devices? And what are your delta phi's right now and are you happy with that?

Harry Atwater:
Our texture is no better than what it was a few years ago: 6 degrees in the plane and 2-3 degrees out of plane.
What is good enough is always application defined. In this application we don't know what is the source of inhomogeneity in electro-optic switching. I also suspect that film roughness is playing a role, because undulations in the film thickness produce electric field variations. Switching of electro-optic materials is a complicated business and I don't know if I can give the answers as to what is good enough.

David Srolovitz:

The question that Bob was referring to was that the inherent angular width of some of these channels is rather broad; and that is something that you are stuck with; it is a materials property. You have some control over that in terms of choice of material and a choice of ions, but it is a material property. Having said that, you can get very different kind of results if you are trying to orient the film via competitive growth versus control of nucleation. But there are other methods that can be quite different and that can used to control alignment and get better and better as the film grows. So for example, and I am not sure if it's quite applicable in IBAD, if you have systems for which the growth of the film is controlled by reactions on the surface such that different orientations, for example a (100) surface grows faster than a (111) surface, directly because of the attachment kinetics, then as you grow the film you should get improvement in orientation as it gets thicker. This is not an ion beam process, but it is one where this limit does not apply.

However, certain things are a materials property, together with the choice of ion.

Xuming Xiong (SuperPower):

I sometimes see TEM cross-sections of IBAD-MgO with seed layer. In these you can see several types of MgO grains. Seed layer is certainly very important. But TEM shows that the MgO grains grow on seed layer, but other grains grow on top of MgO. In fact the second grains grow on top of MgO, and MgO-MgO is a strong bond (presumably not amenable to further alignment). How can you explain the improvement in texture?

Srolovitz:

Let me start but I will then turn it to Bob.

In the simulations that I showed where you see the evolution through the cross section when you see new grains appearing those are strictly a result of the fact that the grains are growing in through the side that you see.

But it is true that if you look at some of the experiments coming out of Bob and Connie's work that there seems to be some sort of re-nucleation as you get thicker. We have never considered that in any of the simulations.

Hammond:

I am not sure if you are referring to the TEM picture that I showed, but my picture is that initially, at least in the "Stanford" mode not the "Caltech" mode as referred to high ion-to-molecule ratios, crystallites of MgO are moving around on the substrate and growing, moving around and still not touching, until they touch and then you have grain boundary minimization. That's my picture.

My one TEM picture was made at 60 Å thickness where there was not complete coverage, just isolated MgO islands which dancing around and are not touching each other.

I think after you get complete coverage then they do grow on top of each other. That's the question in my mind when does the ion beam stop being useful and then you want to go to homoepitaxy. In my case, of the old Stanford result, you want to go up to a thickness which is complete coverage, which may be about 100 Å, and the grain size are cubes of about 100 Å.

Srolovitz:
(Showing a TEM picture) So this is a cross section TEM of the MgO and it looks to me like you have several places where grains re-nucleate. Looks like a patchwork quilt; whether it's a microscopy effect, I don't know.

Hammond:
Remember most of that is homoepitaxy.
In the picture I showed until you get complete coverage things are moving around and the interaction between the "bed" (seed) layer and MgO is very important. If it makes a good bonding you may not get good motion. This is why I chose Si_3N_4 initially which is something I thought would not have strong interaction and would allow for movement.

Matias:
I would like to come back to something that Jim Harper said at the beginning about the technology of ion beams; are the ion beams good enough for the texturing?
One thing that was striking to us initially was that when we (Alp Findikoglu and I) measured the full width of the beam divergence we found that the beam was quite broad; FWHM was 15°, but we were getting texture that was 5°. It was surprising to us that we could obtain that such a good a texture working with this ion beam technology. Then we did an interesting series of experiments where we collimated the ion beam using channel plates. That work probably needs to be revisited in light of new knowledge we now have. The essential conclusion from that work that I took was that it did not matter how much we improved the beam collimation, we were essentially getting the same texture in the MgO. So there is some sort of self-alignment going on here in the material itself, and I am speaking only of the IBAD-MgO. The initial texture may be set by the width of the beam, but beyond that it seems to be determined by the material.

Harper:
That's an interesting comment. You can collimate the beam up to a point.

And Dave (Srolovitz) can probably collimate it best since he puts it in his simulation as a perfectly collimated beam. (Hammond: But it does not help!) But it brings to mind another observation; we've made some AlN deposition, in which case we see (c-axis) tilted texture under some conditions. The width of those textured peaks is really quite narrow compared with the angular range of deposition that we have. We have 3" magnetrons that are not very far from the substrate and there is about a 20° angular range for deposition, and yet the film aligns itself much.

Let's shift blame on to the material and away from the ion source.

We are talking how to get that asymptotic alignment process to "click in" to the fully aligned microstructure. I think of it as a stored energy in the grain boundary misorientations that's going to go away as it gets more and more aligned. So what can we do to make it all go away? I would like to ask this to all the other panelists and anyone else who wants to comment.

Atwater:

Since we were talking about the Caltech way of doing things, I thought I would show just one slide (slide with pictures from *Appl Phys Lett* paper, Brewer et al 2002). This is the dark field TEMs, diffraction patterns, with RHEED patterns at different thicknesses of MgO on SiN, and this was the result that led us to infer that the process of textured grain alignment was a solid phase process, because we were able to deposit a film of 2 nm with no appreciable observation of texture. You can start to see some emerging here in the TEM, you can see the RHEED now is modulated somewhat along the ring, and then finally between 3.7 and 4.6 nm it's absolutely starting to lock in. It's very, very rapid locking in temporally, but of course thickness and time are covarying. So what you can see here are grains that have finally begun to coalesce at 4.8 nm. We essentially see by RHEED a diffraction pattern that is indistinguishable from that of films with considerable amounts of MgO grown on top.

One can infer from this that one limit to the in-plane orientation are the misoriented grains, if this mechanism is one that you believe in. You would want to have some effect for excluding lateral growth and templating effect of misaligned grains.

We tried really hard to do EBSD (electron backscattering diffraction, orientation imaging microscopy) on these grains and it turns out that the grains are too small to do that in any sensible way. I would be interested if anyone else has tried to do this.

Hammond:

Do these grains light up because they have the right orientation? The other ones in the background? Is there still material there?

Atwater:
That's right. By our inference the other material is amorphous. Our interpretation is that the grains grow laterally.

Hammond:
It is my argument, that I gave this morning, that this why you don't get quite as good alignment (in your case)
Under higher ion beam flux things initially islands are isolated from each other and are free to move around. And it is true at Los Alamos too.

Atwater:
Maybe it's better that way, but to me it was significant that this can be a mechanism, that you can have a solid state mechanism. As you know, you can do the same sort of thing with Si. Rafael Reif showed this a long time ago.
You can use an ion beams to preferentially retain seeds in Si. It's a painful process, but I though it was interesting that this same mechanism appears to be operative in MgO, at least in one regime.
So let me propose that if you were really clever about it, and could balance the ion-to-atom ratio so that you could select an extremely low density of seeds, perhaps we could actually laterally (by lateral solid phase growth) suppress growth (nucleation), ie have a low seed density of seed grains and have the overall grain size be considerably larger than it is. But in our experience we were never able to achieve that control or else that phenomenon is not manifest.

Hammond:
We tried that, but we thought that based on Srolovitz's simulations that we were stuck with the channeling orientation of $\pm15°$ and could not get better than that.

Question:
Is it better to have the MgO deposited at 300°C or at 100°C? It isn't obvious from the discussion.

Hammond:
We found that room temperature is better than 300°C. But I think we want to go back and revisit that. I think you may want to have low temperature initially and then high temperature at the final stages as grain alignment occurs.
You would think that you want better mobility, but we found that it grows better at lower temperature.

Slowa Solovyov (BNL):
I would like to make a comment about the measurement of misorientation. We found recently that the scattering power of oriented grains and the scattering power of tilted (defective, geometrically confined small) grains was not the same and not properly taken account of in our standard (rocking curve) measurements. Therefore we were significantly underestimating the amount of misorientation. So the 15° beam producing 5° alignment could be just an artifact based on the assumption that the tilted grains have the same scattering power.

Srolovitz:
I just want to go back to the question of how wide these channels are in terms of their angular width how do we go about getting sharper textures. I just want to make two quick comments, one, in terms of the nucleation picture, the nucleation rates vary very quickly with orientation, up to an exponential, such that even though we are in a broad region of the channeling, we are really centering on one spot. Second observation, both, with regard to Harry's picture where you seem to get nucleation after you have the film, in the film that is already grown, vs yours (Bob's) where you have isolated grains, in both cases you do have a way for grains to interact which is other than just merging. That is when you get a transformation, or you put an island on a substrate you are always going to have long range elastic interactions. And those long range interactions may be oriented in a way that could help correlate the structure over a longer distance. Not having done the calculations this is only speculation.

Xiong:
You calculated this channeling angle using no divergent beam. If you try to take into account a divergent beam what would be the effect?

Srolovitz:
We have not tried it with a divergent beam; My expectation is that it would just make it worse, but we have not tried it. In the simulation you do one ion at a time.

Xiong:
I want to ask Vlad (Matias), you said you used a sharper beam. Do you find that the processing window is wider for a sharper beam?

Matias:
I don't think we can say from our data. We have not looked at that issue systematically. What we did see is similar texture with a very collimated

(couple of degrees) beam of just a few degree divergence and with a broad, 15°, beam.

Question:
On a different topic, in the future what materials systems do you see IBAD benefitting, besides MgO?

Harper:
Are you from a funding agency? (laughter)
Any material where it matters what the orientation of the grains is. And in particular if the material has an anisotropic property. For example, AlN that I mentioned is piezoelectric, and in the tilted c-axis samples that we've made we confirmed that there is an in plane anisotropy in the piezoelectric coefficient, with colleagues at Univ of Conn. So if there is a property that you would like to build into the plane as opposed to perpendicular, like the transducer coefficients. That would be one area.

Atwater:
In general the oxides seem like the right family of materials. For various reasons, because of their anisotropy, anisotropic properties.
I would like to turn it around and ask what likelihood is there do that we think this would be useful for photovoltaics (PV). I have and a lot of people have photovoltaics and solar energy on their minds right now. Is there a viable path forward here?
I have to say that I am personally pessimistic that IBAD will play a role in improving the quality of the photovoltaic materials that I can think of working with, because the constraints for most inorganic semiconductors on the diffusion length essentially that you need in the film are such that you need a few times the thickness, which means the grain size has to be at least a few times the thickness if not more. The experience we had with the oxides and also we did some preliminary work with Si films on MgO to see if we could make Si films that would be suitable for PV, our experience has been discouraging. There are just so many other ways of making films that have longer diffusion lengths or, in the case of poly-Si, ways of mitigating the effects of grain boundaries through hydrogen.
In fact, there are many people now in the PV community that show famously the relative independence of solar cell efficiency on grain size. And the truth is that the smaller the grains (or in the limit amorphous) you have so much hydrogen to passivate deep level defects, that grain size is not necessarily the first thing on everyone's mind in thin film solar cells. I thought I would throw that out there and see what anybody else has to say about that.

Srolovitz:
On the same topic, the question is whether you want to end up with the substrate with a very precise orientation other than the one that nature gives you, versus whether it's the grain size per se and therefore the defect properties of grain boundaries that are important.

In my opinion the materials that are going to be the most interesting are the ones which are further from cubic than even the perovskites. I think Jim mentioned this earlier that materials that are not cubic tend to be very anisotropic, piezoelectrics, ferroelectrics, etc.; all these things that have higher order materials constants.

Alp Findikoglu (LANL):
Having spent a couple of years on the (PV) topic, I can make a couple of comments.
We can start with the basic idea of using a material with the right absorption length comparable to the diffusion length or less, that's the benchmark, to be able to collect the carriers that are generated and then have a nice junction to push them to the right electrode and collect them.
I guess the opportunity that I can see here is in using a Si film on an IBAD template. First of all, I think that it is quite convincing what we have seen is that the majority carrier mobility definitely improves with grain alignment. Grain boundaries for low angle misalignment tend to be less scattering. One part of the question is already answered. The second part which is harder to answer, is what is the effect of grain boundaries on the minority carrier lifetime. Indications are that there is a similar dependence, although we don't have right now a convincing set of results that minority carrier lifetime improves by orders of magnitude by just improving grain alignment. But just one example that we have shows that in a relatively thin film we can have microsecond minority carrier lifetimes. The good thing here is that I think it might be possible to have relatively thin Si films with relatively good grain alignment and combine it with a good light trapping technique, so that the requirement for diffusion length is reduced from hundreds of microns (for wafers) to less than 10 microns for a thin film.

Atwater:
I agree with everything you said, except about the need for grain alignment. Because we have been able to achieve microsecond lifetimes in 2-micron Si films grown by hot-wire CVD with no grain alignment, but with high passivation and we have been able to grow solar cells and so forth.

Findikoglu:
So I would say that the advantage of grain alignment might be the stability of the material and less requirement for further processing in trying to passivate defects at the grain boundaries. We see that as the grain alignment improves, the grain boundaries are already benign and we don't see any effect of passivation on the measured results, at least for the majority carriers.
So this could be a good way to have a simple process that is stable that imitates single crystalline Si properties, but in a platform that uses one tenth of the material.
And with the advantages of using the new techniques that people have been showing recently for light trapping.

IBAD Long Length Application

Mater. Res. Soc. Symp. Proc. Vol. 1150 © 2009 Materials Research Society   1150-RR05-01

IBAD-MgO Architecture at SRL for Long Length IBAD/PLD Coated Conductors

Yutaka Yamada[1], Seiki Miyata[1], Masateru Yoshizumi[1], Hiroyuki Fukushima[1], Akira Ibi[1],
Teruo Izumi[1], Yuh Shiohara[1], Takeharu Kato[2] and Tsukasa Hirayama[2]
[1]Superconductivity Research Laboratory, ISTEC, Tokyo 135-0062, Japan
[2]Japan Fine Ceramics Center, Nagoya 456-8587, Japan

ABSTRACT

Using self epitaxy of a CeO_2 layer, simplified 3 and 4-layer buffer architectures were successfully fabricated for Ion Beam Assist Deposited (IBAD)-MgO template with the structures of $Gd_2Zr_2O_7$ (GZO)/ IBAD-MgO/$LaMnO_3$ (LMO)/CeO_2 and GZO/IBAD-MgO/ CeO_2. Both exhibited high degree of in-plane texturing, around 4 degrees on the CeO_2 layer. Using the 4-layer architecture, a 41 m-long GdBCO coated conductor was successfully fabricated at the production speed of 24 m/h for IBAD-MgO and the critical current recorded high values of 500 to 600 A/cm-width at 77 K. Furthermore, a new IBAD-MgO deposition method was developed using DC-reactive sputtering instead of the ion beam sputtering for MgO deposition and resulted in a high production speed of 150 m/h in a small deposition area (6×20 cm^2).

INTRODUCTION

Among the coated conductor processing, Ion Beam Assisted Deposition (IBAD)-MgO process is promising for the high production speed, which is indispensable for the mass production of the conductor. SuperPower Inc. recently reported km-long production using the IBAD-MgO process with an in-plane texture, $\Delta\varphi$, of 6 to 7 degrees and the high production speed of 120 m/h [1], using an ion beam reactive sputtering IBAD-MgO system with the ion beam gun of $6\times60cm^2$ in area. Compared to the previous IBAD-GZO [2], the IBAD-MgO is textured at a much thinner thickness, about 10 nm [3], and the high production speed of over 100 m/h was realized.

Despite all the success, the architecture of the IBAD-MgO template is still rather complicated. An amorphous Al_2O_3 barrier layer against cation diffusion from the Hastelloy substrate, an IBAD nucleation layer of amorphous Y_2O_3 for the IBAD-MgO and two additional epitaxial layers of homo-epitaxial MgO and $LaMnO_3$ (LMO) layers are used in the recent long tape IBAD-MgO [1]. Meanwhile, for the commercialization of the IBAD-MgO based coated conductor the architecture should be simpler than the above 5 layers in order to reduce cost of the manufacturing equipment. Efforts to reduce the number of buffer layers for IBAD-MgO are reported using LMO or $Sm_xZr_{1-x}O_y$ (SZO) directly deposited on IBAD-MgO [4, 5].

In this study, simplified 3 and 4-layer IBAD-MgO buffer architectures are presented, using two new techniques for a nucleation layer of amorphous GZO and a self-epitaxial CeO_2 layer [6, 7]. Especially, using the 4-layer architecture, we succeeded in development of 1) 41 m long conductor with high I_c of 600 A/cm-width and 2) a new IBAD-MgO system by DC reactive sputtering instead of the previous ion beam sputtering, which exhibited a high production speed of 150 m/h for the IBAD-MgO layer with a small ion beam gun size (6x20 cm^2).

EXPERIMENTAL

On a polished metal substrate an amorphous GZO layer was first deposited at room temperature by a large ion beam sputtering system with a large ion beam gun (6x60 cm^2). Subsequently, an IBAD-MgO layer was deposited on the GZO layer at room temperature by IBAD systems. The assisted ion beam was set with the impinging angle of 45 degrees between the normal to the substrate and the ion beam. During the deposition, the RF power and the Ar$^+$ ion-beam energy of the ion source were controlled and optimized, according to the reported guide line such as the ratio of the etched atom/deposited atom or the ratio of the ion flux/ MgO (or Mg) flux [8]. The details of the deposition conditions were described elsewhere [6]. For an IBAD-MgO long tape fabrication, two different types of Reel-to-Reel IBAD systems were used: 1) a small-sized IBAD system with two 6x20 cm^2 ion beam guns for sputtering and assisting; 2) another small-sized IBAD system with one 6x20 cm^2 ion beam gun for assisting and DC reactive sputtering for MgO deposition. After the IBAD-MgO deposition, an LMO layer was deposited by RF sputtering and then a CeO$_2$ layer was deposited by Pulsed Laser Deposition (PLD) [6]. The deposition temperatures of LMO and CeO$_2$ were varied from 600 to 900 ºC. Lastly, several μm thick GdBCO layer was deposited by the multi-plume and multi-turn (MPMT) and reel-to-reel PLD method [9]. I_c of the coated conductors was measured at 77 K and self-field using the conventional four-probe method with a criterion of 1 μV/cm.

RESULTS and DISCUSSION

4-layer Structures

Using a sputtered amorphous GZO layer, a sputtered LaMnO$_3$ layer and a PLD-CeO$_2$ layer, a 4-layer IBAD-MgO architecture was fabricated as shown in Fig. 1(a). The number of buffer layers was reduced to 4 layers from the conventional 5-layer Al$_2$O$_3$/Y$_2$O$_3$/IBAD-MgO/ epi-MgO/LMO architecture [1]. The first layer of the amorphous GZO layer played a double role of a diffusion barrier against Ni and Cr elements from Hastelloy and a nucleation layer for IBAD-MgO. The 110 nm thick GZO layer effectively suppressed the reduction of the critical

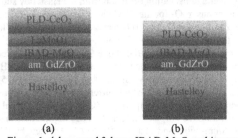

(a)   (b)

Figure 1. 4-layer and 3-layer IBAD-MgO architectures.

temperature, T_c, due to Ni and Cr cation diffusion. The second IBAD-MgO layer was deposited below 10 nm in thickness and formed biaxial grain alignment by an ion beam assisting. The third layer LMO made the following fourth layer of PLD-CeO$_2$ grow stably with the self-epitaxy effect and a high degree of the in-plane texture. In this architecture, 110 nm GZO/ <10 nm IBAD-MgO/ 19 nm sputtered LMO/ 500 nm PLD-CeO$_2$ [6, 7], high degree of in-plane grain alignment of about 3 to 4 degrees was obtained.

3-layer Structures

Recently, we have also developed a 3-layer structure of IBAD-MgO as shown in Fig. 1(b). The LMO layer was further reduced from the above 4-layer structure. The 3-layer structure consisted of 110 nm GZO/ <10 nm IBAD-MgO/ 500 nm PLD-CeO$_2$ [10]. The low value of $\Delta\varphi$ of 3.7 degrees was obtained and now being under investigation for the mechanism of the texturing in spite of the large lattice mismatch between MgO and CeO$_2$. Since the CeO$_2$ layer was formed on the MgO layer with the cube-on-cube relationship, the lattice mismatch is about 28.5% between CeO$_2$ layer and MgO layer as shown in Table I. As reported by Narayan et al. [11], domain matching epitaxy may explain this phenomenon. This kind of epitaxy with a large mismatch was also observed and discussed in the case of SZO [4].

Table I. Lattice mismatch between IBAD-MgO buffer layers. Note that CeO$_2$ layer in 3-layer structure has cube-on-cube relationship in spite of the large mismatch to MgO layer below while CeO$_2$ layer in 4-layer structure has 45 degree rotation relationship to LMO layer below.

4-layer structure			3-layer structure		
Layer and epitaxial direction	Lattice distance, d, (nm)	Mismatch (%)	Layer	Lattice distance, d, (nm)	Mismatch (%)
CeO$_2$(220)	0.1913	1.4	CeO$_2$(200)	0.2705	28.5
LMO(200)	0.1940	7.8	MgO(200)	0.2105	-
MgO(200)	0.2105	-			

IBAD-MgO Buffer Layer Structures Using CeO$_2$ Self-epitaxy

In the above two cases, we have applied PLD-CeO$_2$ self-epitaxy. Figure 2 shows the detailed self-epitaxy effect, which developed improved in-plane grain alignment with increasing thickness. Here, the conventional Al$_2$O$_3$ barrier layer and the Y$_2$O$_3$ nucleation layer were used under MgO layer and 100 nm to 500 nm thick CeO$_2$ layers were deposited on the LMO layer above IBAD-MgO as in Fig. 1(a) [6, 7]. Similarly to the previous case of IBAD-GZO [12], the CeO$_2$ showed improvement in $\Delta\varphi$ from about 40 degrees to around 10 degrees as shown in Fig. 2. Recently, the texturing of IBAD-MgO itself was improved by controlling the ion beam condition and then a better grain alignment of 3 to 4 degrees was obtained after the CeO$_2$ deposition as in the open circle in Fig. 2 [6, 7].

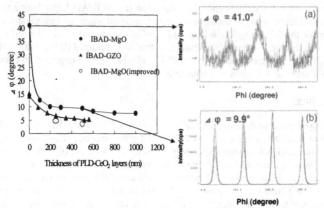

Figure 2. Self-epitaxy effect of PLD-CeO$_2$ on in-plane texturing degree in the IBAD-MgO and the IBAD-GZO cases [12]. The architecture is Al$_2$O$_3$/Y$_2$O$_3$/IBAD-MgO/LaMnO$_3$/CeO$_2$ and IBAD-GZO/CeO$_2$. Recent data (o) on IBAD- MgO showed superior in-plane texturing of CeO$_2$.

Long Tape Fabrication

Using the 4-layer IBAD-MgO architecture in Fig. 1(a), we have fabricated a few pieces of 50 m long IBAD-MgO substrates and obtained mosaic spread in $\Delta\varphi$ of 3.7 degrees [6]. The production speed was 20 m/h for GZO, 24 m/h for IBAD-MgO, 30 m/h for LMO and 5 m/h for CeO$_2$. Subsequently, PLD-GdBCO was deposited by the MPMT-PLD method [9]. Figure 3 shows the transport I_c data along the 41 m length at each 60 cm interval. Almost all parts exhibited high I_c between 500 and 600 A/cm-width except two points. The thickness was 2.5 μm and the J_c value was about 2 MA/cm^2.

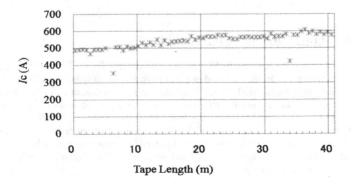

Figure 3. I_c distribution of 41 m long IBAD-MgO/PLD-GdBCO coated conductor.

IBAD-MgO by DC-reactive Sputtering

IBAD-MgO method using DC reactive sputtering and an Mg metal target were also developed [6, 7] instead of the above ion beam sputtering for MgO target because of the high deposition rate peculiar to the reactive sputtering. The same ion beam gun (6x20 cm^2) was used for an assist ion source while the sputtering of Mg was carried out by the DC reactive sputtering. Using this DC reactive sputtering IBAD-MgO system, we have deposited the IBAD-MgO layer at 150 m/h for a 50 m long patch sample, which consisted of 5 pieces of 20 cm long substrates with the GZO nucleation layer connected to 4 pieces of 12 m long Hastelloy tapes. The 5 pieces of 20cm long tape showed $\Delta\varphi$ around 9 degrees after CeO$_2$ deposition and exhibited I_c values of 220 to 286 A/cm-width after GdBCO deposition [6, 7]. Figure 4 shows the TEM microstructure of this sample. On the 4-layer IBAD-MgO buffer layers, PLD-GdBCO was formed uniformly. Furthermore, The MgO layer was estimated about 3 nm in thickness from the detailed microstructure in Fig. 4(b).

The obtained production speed was as high as 150 m/h in spite of the same deposition area as the ion beam sputtering IBAD system (6x20 cm^2 area), whose production speed was 24 m/h. The difference of the production speed, 150 m/h and 24 m/h, was 6 times. Although both samples have the difference of $\Delta\varphi$, 9 degrees for DC reactive sputtering samples and 4 degrees for ion beam sputtering samples, 6 times cannot be explained by the thickness (or deposition time) dependence of $\Delta\varphi$ because the IBAD-MgO thickness (or deposition time) dependence of $\Delta\varphi$ shows about 3 times difference between $\Delta\varphi$ of 9 degrees and 4 degrees according to the detailed IBAD-MgO texturing study [8]. One of the reasons for this large improvement in the production speed is the high MgO deposition rate by the reactive sputtering. From the thickness measurement of IBAD-MgO, about 3 nm, in Fig. 4 (b), the deposition rate of the DC reactive sputtering IBAD-MgO was estimated 12 nm/min. For the ion beam sputtering IBAD-MgO, a smaller value of about 3nm/min was obtained in our ion beam sputtering system and also reported by Arendt et al [13]. Namely, the high production rate of the reactive sputtering is one of the reason for the above high production speed of 150 m/h. Similarly, SuperPower reported the effect of reactive sputtering: a high production speed of 120 m/h for $\Delta\varphi$ of 7 degrees was obtained by using reactive sputtering with 6x60 cm^2 ion gun [1] instead of the conventional ion beam sputtering with the production speed of 65 m/h. This was due to 55% increase in the

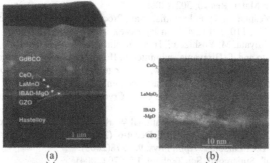

(a)   (b)

Figure 4. TEM images of (a) PLD-GdBCO on the layers GZO/ IBAD-MgO/ LMO/ CeO$_2$ and (b) detailed structure near the IBAD-MgO layer.

deposition rate in the ion beam reactive sputtering system.

In addition to the high deposition rate, the mechanism of IBAD-MgO formation should be considered for the reactive sputtering because Mg atom and oxygen should react during IBAD texturing while MgO mainly deposits in the normal ion beam sputtering using MgO target.

CONCLUSIONS

New simplified 3 and 4 layer IBAD-MgO architectures consisting of GZO/ IBAD-MgO/(LaMnO$_3$)/CeO$_2$ were developed. Notably, PLD-CeO$_2$ exhibited the significant self-epitaxial effect to develop the in-plane grain alignment down to 3 to 4 degrees. Using the 4-layer structure, a 41 m long conductor was produced and high I_c of 500 to 600 A/cm-width was obtained. Furthermore, a new IBAD-MgO process using DC reactive sputtering was developed and a high production speed of 150 m/h was achieved.

ACKNOWLEDGMENTS

This work was supported by New Energy and Industrial Technology Development Organization (NEDO) as the Project for Development of Materials & Power Application of Coated Conductors, M-PACC

REFERENCES

1. V. Selvamanickam, Y. Chen, X. Xiong, Y. Xie X. Zhang, Y. Qiao, J. Reeves, A. Rar, R. Schmidt and K. Lenseth, Physica C. **463-465**, 755(2007).
2. C.P. Wang, K.B. Do, M.R. Beasley, T.H. Geballe and R.H. Hammond, Appl. Phys. Lett.**71**, 2955(1997).
3. A. Ibi, H. Fukushima, R. Kuriki, S. Miyata, K. Takahashi, H. Kobayashi, M. Konishi, T. Watanabe, Y. Yamada, and Y. Shiohara, Physica C **445-448**, 525(2006).
4. L. Stan et al., Supercond. Sci. Technol. **21**, 105023(2008).
5. O. Polat et al., J. Mater. Res **23**, 3021(2008).
6. S. Miyata, M. Yoshizumi, H. Fukushima, et al., Proc. 2008 Spring Domestic Conf. Cryog. Soc. Japan **78**, 109, 110, 111 (2008) (in Japanese).
7. Y. Yamada, S. Miyata, M. Yoshizumi, H. Fukushima, A. Ibi, A. Kionoshita, T. Izumi, Y. Shiohara, T. Kato and T. Hirayama, submitted to IEEE Trans. Appl. Supercond. (2009).
8. V. Matias, US DOE Annual Peer Review, Washington, DC, July 25-27, 2006. http://www.energetics.com/supercon08/archive.html.
9. A. Ibi et al., Proc. 2008 Spring Domestic Conf. Cryog. Soc. Japan **78**, 112(2008) (in Japanese).
10. H. Fukushima et al., presented in International Workshop on Coated Conductors for Applications, CCA2008, Houston, USA, Dec.4-6, (2008).
11. J. Narayan and B.C. Larson, J. Appl. Phys. **93**, 278(2003).
12. Y. Yamada et al., Supercond. Sci. Technol. **17**, S70 (2004).
13. P. N. Arendt et al., Physica C **412- 414**, 795(2004).

Mater. Res. Soc. Symp. Proc. Vol. 1150 © 2009 Materials Research Society   1150-RR05-02

In-Plane Texturing of Buffer Layers by Alternating Beam Assisted Deposition: Large Area and Small Area Applications

Alexander Usoskin and Lutz Kirchhoff

Bruker High Temperature Superconductors GmbH, Siemensstr. 88, D-63755 Alzenau, Germany

ABSTRACT

In standard ion beam assisted deposition (IBAD), the growing film is exposed to inclined ion etching in order to achieve a preferable in-plane orientation of the crystalline structure. Recently, we suggested exposing the film periodically to deposition pulses and to etching pulses, i.e. to assisted beam pulses. As a long sequence of alternations of these two pulses is needed, we named this method "alternating beam assisted deposition" (ABAD). In real application, the substrates exposed to the molecular/atomic flow originating from the sputter source acquire a few nanometer thick layer of yttria-stabilized zirconia (YSZ). In the next step, this layer undergoes ion etching with an Ar ion beam emitted from the source with a particle energy between 200 and 300 eV. Simultaneously with ion-beam exposition, an additional electron beam provides neutralizing of the electrical change in the substrate plane. The ion beam guided 55° at a 55° angle of incidence provides selective etching of the YSZ layer, leading finally after numerous deposition-etching cycles to a sufficiently high quality of in-plane texture in the YSZ layer with the best FWHM values of 8°-9°.

INTRODUCTION

The technology of high-temperature superconductors (HTS) of the second generation requires a biaxially textured template layer onto which the superconductor is deposited. These template buffer layers have to be lattice-matched, chemically compatible, and cost effective in long-lengths processing. At present ion-beam assisted deposition (IBAD) is widely used in the processing of such templates [1-6].

A different method utilizing alternation of the molecular beam and the ion beam, i.e. alternating ion beam assisted deposition (ABAD) [7], was recently suggested in order to improve the quality of the texture.

In this study we describe the background of the ABAD process and also demonstrate different possibilities for ABAD application. One on these is deposition of in-plane textured layers onto wide flexible tape formed as closed loop. Another, more widespread application in template processing of long-length flexible tape is very important in the technology of HTS coated conductors [1-7]. As a special case, we also consider here the ABAD technique used in buffer template deposition across cylindrical substrates. Below we have cited test results which are obtained employing yttria-stabilized zirconia (YSZ) that facilitates a high degree of in-plane texture with characteristic full width at half-maximum (FWHM) of 8-11°.

ABAD

The method we recently introduced in the processing of HTS coated conductors is based on a technique where the template film is periodically exposed to alternating deposition and etching pulses. A schematic view of a set-up developed for realization of this process with flexible substrate tape employed as a substrate 1 is shown in Fig. 1. The tape is guided through the rollers 2a and 2b as a closed loop. Sequentially, each particular area of the tape is pulled through deposition zone D and after this passes through zone E for selective ion etching. All processing elements are installed in the high vacuum (10^{-6}-10^{-7} mbar) chamber that allows high speed (6000 l/sec) pumping in order to enable the operation of ion sources operated under considerable gas flows.

According to this layout, the growing layer undergoes a series of alternating pulses of deposition and ion etching. Such "alternation" is effected in such a way that several nanometers of buffer material are deposited in zone D in each cycle, while in "etching" zone E about 30% of this newly deposited material is removed by ion beam. The ion beam exhibits an incidence angle of 55° which is particularly suitable for yttria-stabilized zirconia template processing. The typical thickness remaining after one deposition / etching cycle corresponds to 2-5 nm.

In practice, the "closed loop" technique described above is utilizable solely for the deposition of wide (20-60mm) tapes. In other cases, especially for long length "narrow" coated conductors, multi-path transport systems similar to those shown in Fig. 2 are used [7]. The transporter depicted in Fig. 2 facilitates multiple treatment of tape in zones D and E with up to 60 alternation cycles. In many cases this number is not sufficient to achieve a high level of in-plane texture. Thus, the tape should be guided through the entire multi-path system several times. This can be done either by employing a closed loop made of the tape or by "reversible" (back and forth) motion of the tape. In the latter case, one of the "D-E" cycles (that corresponds to the moment of motion reverse) is performed in the opposite direction. Inversion of one of the cycles has, nevertheless, some minor influence on the in-plane quality of the final textured buffer layer.

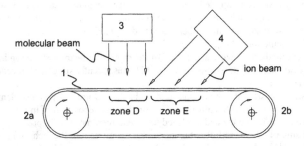

Figure 1. Schematic view of alternating beam assisted deposition (ABAD) employed for processing of in-plane textured template of buffer layer. 1 – substrate tape, 2 – guide rollers, 3 – molecular beam source, 4 – ion beam source (assisted beam). Zones D and E correspond respectively to deposition and selective etching areas.

Figure 2. Multi-path transport system employed in alternating beam assisted deposition (ABAD) onto 4-10mm wide substrate tapes [7].

Further modification of the ABAD technique is depicted in Fig. 3. This modification enables deposition of in-plane textured layers onto cylindrical substrates. This substrate geometry may potentially be used in manufacturing of HTS coils, fault current limiters and field screens. The deposition process follows practically the same procedure as shown in Fig. 1 while the cylindrical substrate may initially be considered as "closed loop". The aperture of the ion beam in this case is more "restricted" by possible variation of the incidence angles, and thus is typically formed as a beam with narrower "in-plane" aperture (i.e. in the cross-section shown in Fig. 3).

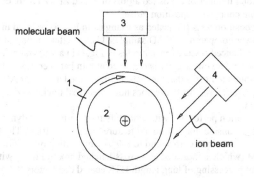

Figure 3. ABAD geometry used for deposition of in-plane textured buffers onto cylindrical substrates. 1 – substrate, 2 – rotating holder, 3 –molecular beam source, 4 – ion beam source.

A further variant of the ABAD technique that can be employed in the case of small discrete substrates which are periodically transported between the D and E areas is already considered in [7].

QUALITY OF ABAD LAYERS; COMPARISON WITH IBAD

Experimentally, we employed yttria-stabilized zirconia (YSZ) as a buffer layer material. Ion etching was performed with an Ar ion beam originating from the Kaufmann source with a particle energy between 200 and 300 eV. Simultaneously with ion-beam exposition, an additional electron beam provides neutralizing of the electrical change in the substrate plane. The ion beam guided at an incidence angle of 55° causes selective etching of the YSZ layer, leading finally, after up to several hundreds deposition-etching cycles, to a sufficiently high level of in-plane texture in the YSZ layer exhibiting the best FWHM values of 8°-9°.

We found that this deposition technique facilitates a better quality of in-plane texture compared to either the "static" etching technique [2] or the IBAD process [1, 3-6]. Essentially, the effect of texture enhancement is achieved in part due to the decrease in substrate temperature because of reduced density in the dissipated energy delivered by two beams. In ABAD case, this energy (or power) is distributed within two zones instead of the single one used in IBAD [3], [4]. A further physical mechanism that may be responsible for the formation of highly textured template is an enhancement of the texture level due to multiple stages of homoepitaxial film growth which follows after each selective etching step. This should result in the formation of sufficient texture quality at earlier stages of film growth. This latter effect is experimentally observed as shown in Fig. 4 below.

Fig. 4 illustrates a thickness dependence of in-plane texture quality which was achieved in the course of ABAD optimization. Solid curve and squares correspond to state of the art of IBAD processing of bi-axially textured YSZ [8].

There were three steps of ABAD optimization with regard to deposition speed, substrate temperature and ratio, η, of the number of Ar ions provided by assisted beam per cm^2 during one etching pulse to the total number of deposited atoms of Y and Zr which are condensed onto $1 cm^2$ substrate surface in the course of deposition pulse.

First set of deposition tests (indicated as triangles in Fig. 4) resulted in moderate texture quality which was lower compared to IBAD samples. Using better cooling of samples during ABAD we achieved enhanced texture (open circles in Fig.4) in a wide range of η values, i.e. for η from 1.2 to 1.9. The best texture quality is observed at η in between 1.5 and 1.9. Dashed curve connecting the data points indicates that the tests are performed at similar ABAD conditions except variation of η. Therefore, this curve does not reflect real thickness dependence of FWHM while η is not constant.

Further optimization performed with smoothly transported and "dynamically" cooled tape substrate leaded to considerably improved texture level of 10-10.5°. The best FWHM values achieved here with this technique correspond to 8-9°. Combination of optimized ranges of layer thickness and FWHM (which is indicated as a cross-hatched box in Fig. 4) with $\eta = 1.6$-1.8 is employed presently in processing of long length HTS coated conductors (CC) with lengths from

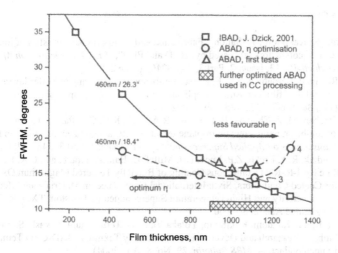

Figure 4. Quality of in-plane texture achieved in the course of ABAD optimization employing small test samples. Curve and squares correspond to state of the art of IBAD processing of bi-axially textured YSZ [8]. Circles 1-4 correspond to η = 1.2, 1.5, 1.9 and 2.8 respectively, where η denotes ratio of the number of Ar ions provided by assisted beam per cm^2 during one etching pulse to the entire number of deposited atoms of Y and Zr which are condensed onto $1cm^2$ substrate surface in the course of deposition pulse.

40 to 100 m [9]. In this case, ABAD allows shortening the entire deposition time by 25-35% compared to the time needed in case of IBAD process. Similarly, this benefit may be quoted in terms of enhanced texture quality at constant deposition time which is roughly proportional to the layer thickness.

SHORT CONCLUSION AND OUTLOOK

Thus, we have shown that the alternating beam assisted deposition (ABAD) is a beneficial processing tool which facilitates not only the enhancement of in-plane texture but also enables an additional degree of freedom in the construction of processing machines where deposition and ion etching zones may be decoupled, i.e. divided into two practically independent processing sectors. A wide range of parameters is found where faster texture formation takes place in comparison to ion beam assisted deposition (IBAD).

New features of the ABAD processing are already implemented in construction of new ABAD coating machine that should exhibit enhanced performance particularly regarding the processing throughput.

REFERENCES

1. Y. Iijima, K. Kakimoto, T. Saitoh, "Fabrication and transport characteristics of long length Y-123 coated conductors processed by IBAD and PLD", *IEEE Transactions on Applied Superconductivity*, **13**, Issue: 2, Part 3, 2466-2469 (June 2003).
2. R. P. Reade, P. Berdahl, and R. E. Russo, Ion-beam nanotexturing of buffer layers for near-single-crystal thin-film deposition: Application to $YBa_2Cu_3O_{7-\delta}$ superconducting films, *Appl. Phys. Lett.*, **80**, No. 8, 1352-1354 (Feb 2002).
3. T. G. Truchan, M. P. Chudzik, B. L. Fisher, R.A. Erck, K.C. Goretta, U. Balachandran, Effect of ion-beam parameters on in-plane texture of yttria-stabilized zirconia thin films., *IEEE Transactions on Applied Superconductivity*, **11**, Issue: 1, Part 3, 3485-3488 (2001).
4. M.P. Chudzik, R.A. Erckb, Z.P. Luoc, D.J. Millef, U. Balachandranb, and C.R. Kannewurf, High-Rate Reel-to-Reel Continuous Coating of Biaxially Textured Magnesium Oxide Thin Films for Coated Conductors, Sixth International Conference on Materials and Mechanisms of Superconductivity and High-Temperature Superconductors, Houston, TX, Feb. 20-25, Book of Extended Abstracts, 2000.
5. Yasuhiro Iijima, Kazuomi Kakimoto, Yutaka Yamada, Teruo Izumi, Takashi Saitoh, and Yuh Shiohara, Research and Development of Biaxially Textured IBAD-GZO Templates for Coated Superconductors, *MRS Bulletin*, **29**, No. 8 (Aug 2004).
6. Development of long Y-123 coated conductors for coil-applications by IBAD/PLD method Y. Iijima, K. Kakimoto, Y. Sutoh, S. Ajimura, T. Saitoh, *IEEE Trans. on Appl. Supercond.* **15**, Issue 2, 2590 – 2595 (2005).
7. A. Usoskin, L. Kirchhoff, J. Knoke, B. Prause, A. Rutt, D. E. Farrell, Processing of Long-Length YBCO Coated Conductors Based on Stainless Steel Tapes, *IEEE Trans. on Appl. Supercond.*, **17**, Issue 2, No. 2, 3235 - 3238 (2007).
8. Juergen Dzick, Mechanismen der Ionenstrahlunterstuetzten Texturbildung in Yttrium-Stabilisierten Zirkondioxid-Filmen, Dissertation, Georg-August-Universitaet zu Goettingen, 2001, pp. 72-73.
9. A. Usoskin, L. Kirchhoff, J. Knoke, et al., Processing of Long-Length YBCO Coated Conductors Based on Stainless Steel Tapes, *IEEE Trans. Appl. Supercond.*, **17**, no. 2, 3235-3238 (Jun 2007).

Mater. Res. Soc. Symp. Proc. Vol. 1150 © 2009 Materials Research Society   1150-RR05-04

Xuming Xiong, Karol Zdun, Sungjin Kim, Andrei Rar, Senthil Sambandam, Robert Schmidt, Yimin Chen, Kenneth Lenseth and Venkat Selvamanickam

SuperPower, Inc., 450 Duane Ave, Schenectady, NY 12304 USA

ABSTRACT

SuperPower's IBAD (ion beam assisted deposition) MgO process has been changed to reactive ion beam sputtering of Mg metal target instead of MgO ceramic target, which gives ~ 60 % increase in deposition rate. The process speed was increased from 195 m/h[*] to 360 m/h. Texture of IBAD MgO tapes by this reactive process was found to degrade faster during a long length run. This uniformity issue was resolved with feedback control during long runs and from run to run. The bottleneck of the alumina layer process due to slow ion beam sputtering deposition was removed by new high rate reactive magnetron sputtering at transition mode. The new alumina process speed can be as high as 3,000 m/h in our Pilot Buffer system. The yttria layer process was also changed to high rate reactive magnetron sputtering at transition mode with achievable speed as high as 10,000 m/h in our Pilot Buffer system. The routine production speed of alumina and yttria is 750 m/h due to a limitation of the tape driving system. The high rate magnetron-sputtered alumina/yttria yields the same texture of IBAD MgO as ion beam sputtered alumina/yttria. Now we are routinely producing IBAD MgO template tapes of ~ 1.4 km with a uniform in-plane texture ~ 6-7 degrees. A record high critical current of 813 A/cm over a one meter length and a world record critical current times length value of > 233,810A-m was obtained with our routinely-produced high throughput IBAD MgO buffers. The requirements for a better IBAD texturing layer than IBAD MgO are also suggested.

1. INTRODUCTION

SuperPower Inc is developing and manufacturing HTS (high temperature superconductor) coated conductors for commercial use. Challenges for HTS coated conductors to become commercially viable are long length, high quality, and low cost. High throughput processing is a must for cost reduction and production capacity to meet market needs. SuperPower uses an IBAD MgO approach [1-4] to make HTS coated conductor tape. An IBAD MgO or IBAD MgO-like process is the most promising high throughput, high quality process to meet the market requirements. The in-plane texture of our routinely-manufactured HTS wire is between 2-3 degrees. Such an excellent texture opens the way for extremely high I_c (critical current) which is a very effective way to reduce cost and increase wire performance. Another advantage of the IBAD MgO process is the flexibility to choose different substrates to meet different requirements of cost, processes and applications. For example, by choosing a thin and flexible substrate, this type of HTS wire can greatly enhance the engineering current density for high-field coil applications. The separation of the barrier layer and the texturing layer in the IBAD MgO-like process is also advantageous since it allows the flexibility of choosing different

---

[*] All speeds in this paper are the equivalent speed of 4mm wide tape

barrier buffers to meet different requirements of cost, process and applications. For example, alumina has very good chemical stability and anti-diffusion ability, but with a usual hexagonal crystal structure, it is difficult to use as a buffer for rolling assisted biaxially textured substrates (RABiTS). It can, however, be used as an IBAD MgO barrier layer.

We transitioned from IBAD YSZ technology [5] to IBAD MgO technology with support from Los Alamos National Laboratory (LANL) and made a substantial improvement in the IBAD buffer process speed up to 195 m/h [4]. Since then, we further improved our process in order to meet our milestones for commercialization of HTS wire. In this paper, we report our progress in the high throughput process of producing kilometer length, high quality, and low cost IBAD MgO buffer. The routine buffer process speeds have been increased to 340 m/h – 750 m/h, and the routine process length was increased to ~ 1.4 km. Process steps were reduced from 5 steps to 3 steps. All of these changes have resulted in about a four-fold increase in throughput. All of these achievements were made in our Pilot IBAD and Pilot Buffer systems [4] without new capital investment.

2. Choice of High Throughput, High Yield IBAD MgO Buffer Structure

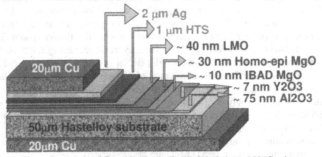

Fig.1. Structure of SuperPower's IBAD MgO-based HTS wire

It is critically important to identify, evaluate, and choose a high yield, robust process for production. Among the numerous processes reported in the literature, some are good for lab demonstration or academic interest. Further, yield-throughput study and process control technique development are needed to see if it can be a practical production process.

Figure 1 displays the present architecture of second-generation (2G) HTS wire made at SuperPower, Inc. It includes 5 buffer layers between the metal substrate and HTS layer in order to make it robust during our MOCVD HTS film coating. This buffer structure is similar to the buffer structure LANL transferred to us, except for the LaMnO$_3$ (LMO) cap layer. The reason for our change to LMO was presented in a previous paper [4].

The first layer, which mainly serves as a diffusion barrier, is alumina, due to its good thermal and chemical stability, excellent anti-diffusion ability and low cost. The cost of Al is at least 20 times lower compared with rare earth metals. Al-Y-O [6] and GZO [7] have been reported as replacements for the alumina plus seed layer. Alumina is still our best option at this time with the combined considerations of performance, yield, cost, easy processing, and safety, etc. The alumina layer was previously processed in the Pilot IBAD system by ion beam sputtering at a speed of 120 m/h. It had been the bottleneck in our total buffer throughput, as it required ~ 40 hours to coat one 1.4 km Al$_2$O$_3$ tape. We increased the process speed to 195 m/h by expanding the deposition zone to the limit of the Pilot IBAD system, but it was still a bottleneck after we increased process speeds of IBAD MgO, HE-MgO (home epitaxial MgO), and LMO to 360 m/h. Hence, we needed to find a new way to break through this bottleneck without new capital investment and a long time period from design to full qualification for routine production. We accomplished this goal with a reactive magnetron sputtered Al_2O_3 process working at transition mode to achieve a high deposition rate of ~ 5 nm/s. This deposition rate is high enough to reach a 3000 m/h process speed with our Pilot Buffer system. Our magnetron sputtered Al_2O_3 film has no problem as a barrier layer for IBAD MgO-MOCVD HTS tape, and yielded the same texture and I_c as that of the ion beam sputtered Al_2O_3 layer. Both O_2 partial pressure and bias voltage are well controlled through the production of long tapes. Fig.2a and Fig. 2b show the good process stability of our magnetron sputtered alumina process. Further, the IBAD MgO process condition needed no re-tuning with our magnetron sputtered Al_2O_3. The alumina thickness is ~ 70 nm in our routine production run, but several of our test tapes with ~ 40 nm alumina showed no degradation of I_c, thus indicating that ~ 40 nm thick alumina is enough as a barrier.

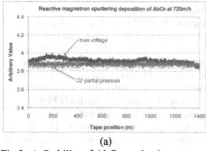

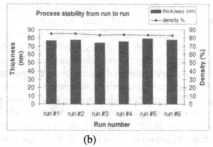

(a)   (b)

Fig.2. a). Stability of Al_2O_3 production run,   b). Al_2O_3 process stability from run to run

The second layer is the seed layer, providing a suitable surface for IBAD MgO texturing. An amorphous Si_3N_4 layer was used as the seed layer during the invention of the IBAD MgO process [1]. It is claimed in the Stanford work that this seed layer has to be amorphous. Later LANL developed yttria as a seed layer with a larger process window and better chemical stability in an oxidation environment at high temperature; It has been reported by LANL also that the seed layer is nano-crystalline [8].

SuperPower is still using yttria as the seed layer in consideration of the process window, long term process stability and cost. A high rate Y_2O_3 process with reactive magnetron sputtering at transition mode was developed with a speed > 10,000 m/h in our Pilot Buffer system. But Al_2O_3 and Y_2O_3 routine production speeds at present are 750 m/h due to a limitation in the tape driving system.

The third layer is the key layer to form biaxial texture. The IBAD MgO approach is still the best choice right now. Two shortcomings of MgO we find are the large lattice mismatch with HTS material and the difficulty to achieve long time reliable reactive sputtering in transition mode. Is there better IBAD material than IBAD MgO? We suggest the following three screening requirements for research toward a better IBAD material: 1) high enough mobility to form good texture during the nucleation stage; 2) the tendency to form small grains after nucleation for easy alignment by ion beam This usually requires strong atom-atom bonding material, and 3) good thermal/oxidation stability in order to withstand later buffer and HTS coating processes. Also, (001) out-of-plane orientation is preferred for coated conductor application. Other orientations like (111) may be preferred for other applications like ferroelectric devices. The co-existence of these requirements can possibly be found in a material with both large bonding energy and a strong ionic bond. The large bonding energy helps to form small grains; and a strong ionic bonding helps the surface mobility since ionic bonding has no direction. The oxides of Group I and Group II metals in the periodic table normally have high degree ionic bonding. The oxides of Group I are not practical for production due to their high activity in air. They also have relatively lower melting points and antifluorite structure. Oxides of Group II are also not stable in air except for BeO, MgO and possibly CaO. BeO is not a very good ionic bonding material due its relatively higher electronegativity. CaO reacts much more quickly with CO_2 and H_2O, so it is not very stable in air without protection. If a protection layer can be deposited on IBAD CaO in vacuum immediately after IBAD CaO, it may be able to be used in HTS wire, and it is possible that the texture of IBAD CaO [6] can be further improved by using a seed layer other than yttria.

Our IBAD MgO process speed was increased to 360 m/h by switching to reactive ion beam sputtering of a Mg metal target. Reactive sputtering of the Mg target gives a higher (60 %) deposition rate compared with an MgO ceramic target, and also enables a four-fold reduction in target cost. We obtained very good texture and I_c with short (~ 5 meters) IBAD tapes processed at 360 m/h during optimization runs, but when we started to process long tapes with this new process, we experienced serious texture degradation with tape lengths as shown in Fig. 3a. The deposition rate of reactive sputtering is very sensitive to very small shifts like temperature and partial pressure of residual impure gases during processing. Additionally, the IBAD MgO process is also a sensitive process which results in a large change in texture with just a few percentage points change in some key process parameters such as ion to atom ratio. Fig. 3b shows examples of a less than 10 % change in deposition rate (so the ion to atom ratio change as ion current is kept the same) leading to a more than 30 % change in texture. So feedback control has to be used to achieve stable long length production. By implementing feedback control, we are able to achieve good stability over long lengths with this reactive IBAD MgO process. Fig. 4 shows examples of process stability of our reactive IBAD MgO process at a speed of 360 m/h.

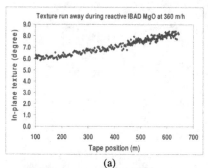
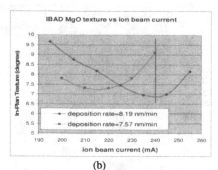

(a)   (b)

Fig. 3a. IBAD MgO texture degradation in our initial run; Texture measured with ~ 30 nm homo-epitaxial MgO and ~ 30 nm LMO

Fig. 3b. IBAD MgO texture change vs the assisting ion beam current change. Texture measured with ~ 30 nm homo-epitaxial MgO and ~ 30 nm LMO.

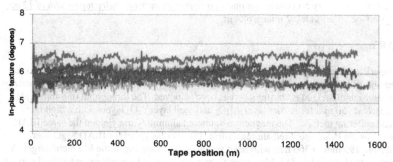

Fig. 4. Examples of IBAD MgO process stability. Texture measured with ~ 30nm LMO and ~30nm homo-epitaxial MgO

A record high critical current of 813 A/cm over a one meter tape and a world record critical current times length value of > 233,810A-m were obtained with our routinely-produced high throughput IBAD MgO buffers. Fig. 5 shows our world record of the current times length metric achieved in August 2008.

127

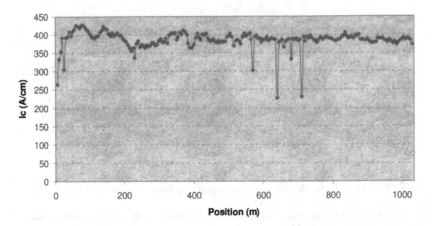

Fig. 5. I_c of superconductor deposited on one of our routinely produced IBAD MgO buffer tapes. I_c measured at 77 K, every 5 m using *continuous dc currents* *over the entire tape width of 12 mm* *(not slit).* Voltage criterion = 0.2 microvolt/cm

SUMMARY

SuperPower's choice of an IBAD MgO buffer structure was discussed with a focus on seed layer and IBAD texturing layer. High throughput, high yield, routine production of greater than kilometer long IBAD MgO buffer tape has been achieved. The throughput of 4 mm-wide wire has reached 750 m/h for barrier layer Al_2O_3 and seed layer Y_2O_3 layers with a high rate magnetron sputtering method. The magnetron-sputtered alumina/yttria yielded the same IBAD MgO texture as ion beam sputtered alumina/yttria. The process speed for IBAD MgO was increased from 195 m/h to 360 m/h with reactive ion beam sputtering of the Mg metal target. A record high critical current of 813 A/cm over a one meter tape and a world record critical current times length value of > 233,810 A-m were obtained with our routinely-produced high throughput IBAD MgO..

ACKNOWLEDGMENTS

This work was partially supported by the Title III office, U.S. Department of Energy and the Air Force Research Laboratory.

REFERENCES

1. C. Wang, K. Do, M. Beasley, T. Geballe, R. Hammond, *Appl. Phys. Lett.*, **71**, 2955(1997)
2. J. Groves, P. Arendt, S. Foltyn, Q. Jia, T. Holesinger, L. Emmert, R. DePaula, P. Dowden and L. Stan,. *IEEE Trans. Applied Supercond.*, **12**. 2651(2003).

3. V. Matias, B. Jibbons, A. Findikoglu, P. Dowden, J. Sullard, and J. Coulter, *IEEE Trans. Applied Supercond.,* **15**, 2735(2005)
4. X. Xiong, K. Lenseth, J. Reeves, A. Rar, Y. Qiao, R. Schmidt, Y. Chen, Y. Li, Y. Xie, and V. Selvamanickam, *IEEE Trans. Applied Supercond.,* **17**, 3375(2007).
5. Y. Iijima, K. Matsumoto, *Superconductor Science & Technology,* **13**, 68(2000)
6. V. Matia, Q. Jia, *DOE Superconductivity for Electric Systems 2008 Annual Peer Review,* July 29-31, 2008, Arlington, VA, USA.
   http://www.energetics.com/supercon08/agenda.html
7. Yutaka Yamada, Seiki Miyata, Masateru Yoshizumi, Hiroyuki Fukushima, Akira Ibi, Akio Kionoshita, Teruo Izumi, Yuh Shiohara, Takeharu Kato and Tsukasa Hirayama, 2008 Applied Superconductivity Conference, presentation 3MA03, Chicago, Illinois USA, August 17 - 22, 2008
8. Paul Arendt, Steve Foltyn, *DOE Superconductivity for Electric Systems 2003 Annual Peer Review,* July 23-25, 2003, Washington, DC, USA.
   http://www.energetics.com/supercon03.html

Texturing by Other Techniques

Mater. Res. Soc. Symp. Proc. Vol. 1150 © 2009 Materials Research Society   1150-RR06-03

Abnormal Grain Growth Behavior in Nanostructured Al Thin Films on SiO₂/Si Substrates

Flavia P. Luce[1], Paulo F. P. Fichtner[2], Luiz F. Schelp[3] and Fernando C. Zawislak[1]

[1]Instituo de Física, Universidade Federal do Rio Grande do Sul, 91501-970 Porto Alegre, RS, Brazil
[2]Escola de Engenharia, Universidade Federal do Rio Grande do Sul 91501-970 Porto Alegre, RS, Brazil
[3]Departamento de Física, Universidade Federal de Santa Maria, 97105-900 Santa Maria, RS, Brazil

ABSTRACT

We report on the formation of nanocrystalline Al thin films (180 nm thick) via magnetron sputtering technique using a step-wise deposition concept where columnar growth is inhibited, giving place to the development of a nanocrystalline mosaic grain arrangement with characteristic diameters of ≈30 nm and small size dispersion. The thermal evolution of the grain size distributions is investigated by transmission electron microscopy (TEM) in samples annealed in high vacuum for 3600 s. For the temperature range $300 \leq T \leq 462$ °C the system presents a 3-D regular growth behavior up to sizes ≈70 nm. For T = 475 °C a rather sharp transition from normal to abnormal grain growth occurs. The grains extend to the film thickness and present mean lateral dimensions of ≈1000 nm. The observed phenomenon is discussed in terms of a synergetic grain boundary mobility effect caused by the characteristics of the initial nanogranular grain boundary morphology.

INTRODUCTION

Thin film structural features such as grain size, size distribution, texture and morphology can significantly influence the desired performance in many applications. For example, metallic interconnections of microelectronic devices with large, defect free grains are typically more resistant against electromigration failures than small grain size counterparts [1]. In contrast, the formation of nanosized grains presenting a monomodal size distribution and rather small size dispersion can improve the wear resistance of hard coatings [2]. In view of such wide variety of configuration requirements, there have been significant efforts to understand and model the grain formation and growth micromechanisms.

Usually the deposited Al films have a columnar grain structure. During annealing treatments in most cases the grain size follows a monomodal distribution. However it has been observed that in some situations the coarsening can take place via abnormal grain growth [3]. Texturization and the release of strain energy are considerate important causes of abnormal grain growth [4].

In the present contribution we report on the formation of nanocrystalline Al thin films via magnetron sputtering technique using a distinct deposition concept, where columnar growth is inhibited and gives place to the development of a mosaic-like nanocrystalline grain arrangement with small size dispersion. The thermal evolution of the grain size distribution is investigated by transmission electron microscopy (TEM). As opposed to the usual concepts, in this particular structure arrangement a rather sharp transition from normal to abnormal grain growth occurs within a temperature interval of $\Delta T \approx 13$ °C, where average grain size increase by a factor larger than 20. The observed phenomenon is discussed in terms of a synergetic grain boundary mobility effect caused by specific characteristics of the nanogranular grain boundary morphology. This leads to predictability and stability criterions for the normal to abnormal growth transition, thus stimulating distinct applications possibilities.

EXPERIMENT

The aluminum films were deposited onto 130 nm thick SiO_2/Si films. Two types of Al films have been produced via magnetron sputtering. The first one, referred as columnar or bamboo-like film, was obtained by the standard continuous deposition process to a thickness of 385 nm with a base pressure of ≈ 0.3 μPa. The second type, here referred as mosaic-like, was obtained by using a step-wise deposition method. The main difference with respect to the columnar film is that between each layer of Al (30 to 40 nm) and the next, the deposition stops for a time necessary to form a monolayer of Al oxide. The time depends on the base pressure, which in this case was ≈ 0.1 μPa. Figure 1 shows diagrams illustrating the structure of both films after deposition. The thickness of mosaic-like film was of ≈ 180 nm.

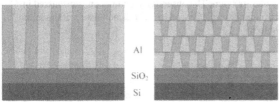

Figure 1. Schematic diagram of the columnar-like and the mosaic-like film deposited by magnetron sputtering.

Some films were annealed in high vacuum (1 mPa) for periods of 1 hour within the temperature range from 300 to 500°C. As deposited and annealed samples were investigated by TEM, performed at 200 kV using specimens prepared by mechanical polishing followed by ion milling.

RESULTS AND DISCUSSION

Figure 2 shows TEM micrographs from columnar and mosaic-like films, either as-deposited or annealed for 1 hour at 500 °C in high vacuum. Figures 2(a) and 2(b) show the sample prepared under continuous deposition growth. They present a columnar morphology with mean width of ≈80 nm that do not change upon thermal annealing. The behavior of the mosaic-like film under the same annealing conditions is completely different. The as-deposited sample, Figure 2(c), contains nanograins with mean sizes of ≈30 nm. After annealing, the grains acquire the thickness of the film and present lateral sizes ranging from 0.4 to 1.5 μm (Figure 2(d)).

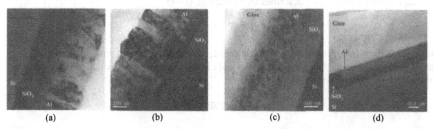

(a)   (b)   (c)   (d)

Figure 2. Cross-section TEM micrographs showing: (a) columnar-like film as deposited and (b) annealed at 500 °C for 1 hour, (c) mosaic-like film as-deposited and (d) annealed at 500 °C for 1 hour.

Figure 3 presents cross-section TEM micrographs showing the grain arrangement obtained by the discontinuous deposition process (mosaic-like film) followed by post deposition thermal treatments for 1 hour at temperatures of 450, 462 and 475 °C. These correspond to the limit temperatures for three characteristic grain morphology regimens. For temperatures up to 450 °C the grains present a mosaic-like arrangement preserving the layered structure from the deposition process. Their growth behavior is rather slow and occurs in both horizontal and vertical directions, being the vertical direction defined as perpendicular to the sample surface.

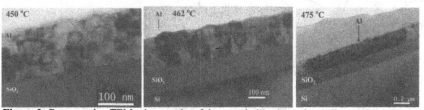

Figure 3. Cross-section TEM micrographs of the mosaic-like films thermally treated at the indicated temperatures.

For T ≈ 462 °C (i.e. 450 °C < T < 475 °C), preferential grain growth occurs within the vertical direction, thus leading towards a columnar morphology. For T ≥ 475 °C the grains grow to a vertical size corresponding to the film thickness and to horizontal dimensions of the order of 1 μm. The conspicuous grain growth behavior observed for T ≥ 475 °C can be characterized by a rather rapid horizontal displacement of somewhat vertically oriented grain boundaries. The average minimum grain boundary velocities correspond to 14 nm/min while the velocities for T = 462 °C correspond to 0,26 nm/min.

Normal grain growth

From 300 to 462 °C the average size of the grains increases from 30 nm to ≈70 nm. In this region the growth is well described by a parabolic law [5]

$$D^2 = D_0^2 + Kt, \qquad (1)$$

where D is the mean grain diameter of each sample annealed at different temperatures, D_0 is the mean grain diameter of the as-deposited film and t is the annealing time, which in this work was always 1 hour. The parameter K is written as

$$K = (K'/T)\exp(-\Delta G^a/kT), \qquad (2)$$

where K' is a constant related with the interface energy density γ. The exponential term and the T^{-1} dependence are related with the mobility M of the grain boundary. Equation (1) was used to fit the experimental data for the evolution of the grain growth observed for each annealing temperatures up to 462 °C. This is plotted in Figure 4(a). The experimental results are well fitted by the parabolic law, thus indicating that, from 300 to 462 °C there is a normal grain growth behavior.

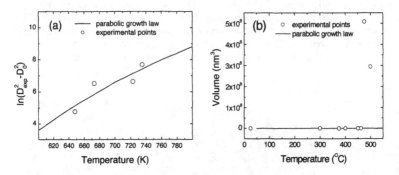

Figure 4. (a) Fitting of the experimental data up to 462 °C using the parabolic law (full line). (b) Volume of the grains as a function of annealing temperature. The data for 475 and 500 °C are completely out of range of any parabolic law fit.

Abnormal grain growth

However, Fig. 4(b) also demonstrates that the experimental grain growth evolution of films annealed at 475 and 500 °C not only deviate significantly but also cannot respond to a parabolic law. The variation in grain volume was from $\approx 3 \times 10^2$ nm^3 at 462 °C to $\approx 5 \times 10^8$ nm^3 at 475 °C. This rapid and abrupt grain growth can only be explained in terms of a significant increase of grain boundary mobility. The grain boundary velocity behavior can be expressed as

$$v = (C\Delta G/TV_m)exp(-\Delta G^a/RT),\qquad(3)$$

where C is a constant, R is the gas constant, T is the annealing temperature, ΔG is the difference in free energy between neighbor grains, ΔG^a is the activation energy for atomic migration between grains and V_m is the molar volume. The most important term in Equation (3) is ΔG. The difference in free energy between neighbor grains acts as a driving force indicating that some special grains have more favorable conditions to grow. In the present work we assume that ΔG is significantly influenced by the grain interface morphology. The columnar films are characterized by rather flat interfaces which can lead to very small ΔG or to a system with 120° equilibrated surface tensions [3], thus inhibiting grain growth even for treatments at 500 °C. In contrast, although some grains from the mosaic films indeed acquire a near columnar morphology after thermal annealing at 462 °C, they may have grain neighbors of small dimensions or present rather sinuous interface segments with curvature radius of the order of 10 to 50 nm, consistently with the characteristics of their preceded morphology. Grain curvatures within such a nanometric scale may significantly enhance grain boundary migration velocities [6, 7]. In the present case, following the concepts from references [6, 7], we assume that each curved grain boundary segment may indeed migrate quite rapidly. As a given grain boundary segment migrates, it may affect the neighbor ones producing a synergetic behavior which ultimately results into a rapid migration of the total interface. Hence, once a fast growth process starts for certain specific grains, it leads to the observed abnormal grain growth process reported here.

CONCLUSIONS

The columnar-like film did not change its structure even after 1 hour annealing at 500 °C. This behavior can be understood as a consequence of their rather planar grain boundaries. A completely different result was observed for the mosaic-like films. It was possible to identify two distinct regimes controlling their grain growth. In the range of temperatures from as-deposited (i.e. room temperature) to 462 °C the growth process follows a rather normal behavior obeying a parabolic law. In this region the size of the grains vary from ≈ 30 nm in the as-deposited sample to ≈ 70 nm in the sample treated at 462 °C. For annealing temperatures of 475 and 500 °C the grains grow abruptly and very rapidly, expanding to the total film thickness and presenting a lateral growth with a mean size of ≈ 1000 nm. This behavior can be understood in terms of an enhanced grain boundary mobility caused by nanometric sinuous boundary segments. We suggest that the movement of each specific segment may affect the neighbor ones enhancing synergistically the overall grain boundary mobility, thus resulting in a rather abrupt abnormal grain growth process.

ACKNOWLEDGMENTS
This work was supported by the Brazilian Agencies FINEP, FAPERGS, CNPq and CAPES.

REFERENCES

1. K. N. Tu; J. Appl. Phys. 94 (2003) 5451.
2. S. Veprek, M. G. J. Veprek-Heijman, P. Karvankova, J. Prochazka; Thin Solid Films, 476 (2005) 1.
3. D. A. Porter, K. E. Easterling; *Phase Transformation in Metals and Alloys*, 2 ed., London: Chapman & Hall, 1992, pp. 110-184.
4. F. Ma, J-M. Zhang, K-W. Xu; Appl. Surf. Sci. 242 (2005) 55.
5. P. R. Rios; Scripta Mater. 40, 6 (1999) 665.
6. M. Upmanyu, D. J. Srolovitz, L. S. Shvindlerman and G. Gottstein, Acta Matererialia 47 (1999) 3901.
7. L. Zhou, H. Zhang, D. J. Srolovitz, Acta Materialia 53 (2005) 5273.

Mater. Res. Soc. Symp. Proc. Vol. 1150 © 2009 Materials Research Society   1150-RR06-05

Artificial Grain Alignment of Organic Crystalline Thin Films

Toshihiro Shimada

Department of Chemistry, The University of Tokyo, Bunkyo-ku, Tokyo 113-0033, Japan

ABSTRACT

It is important to obtain single crystalline organic thin films for electronics and optics applications. Due to the mismatching in the crystal symmetry, it is difficult to align the crystalline grains of organic molecular films even on single crystalline surfaces. We have developed several techniques for the artificial grain alignment in organic epitaxial growth. (1) Use of nanoscale-textured surfaces prepared by step bunching of vicinally-cut single crystalline surfaces, in which the height of the steps is critically important. (2) Application of external electric field. (3) Optical excitation of the molecules which can be applied to the polar semiconducting molecules. These techniques might be applicable to other materials including ionic materials and ferroelectrics.

INTRODUCTION

For the application of the organic materials to electronics and photonics, the crystallinity and the alignment of grains will become more and more important. It is difficult to achieve highly ordered grains of organic crystals because the crystal structure of the organic materials has low symmetry and because there are many crystal polymorphs with similar thermodynamic stability. There are many reports of epitaxial growth of organic molecules on inorganic single crystals but it is not enough in most cases. Re-evaporation of the films on organic crystals with low cohesive energy is also a great problem because the annealing process in vacuum for the grain evolution, which is frequently used for the inorganic materials, cannot be used for the organic materials.

In this paper, we summarize the various attempts for the artificial grain alignment of organic thin films focusing on the results of our groups. The first technique is the use of the nanoscale-textured surfaces prepared by step bunching of vicinally-cut single crystalline surfaces. The height of the steps is critically important: it must be greater than the height of one molecular layer. The mechanism of the grain alignment cannot be explained only by thermodynamic considerations, and kinetic effects revealed by molecular beam experiments must be involved. The second technique is the application of external electric field. We found that there is threshold field strength for the grain alignment. The third one is the optical excitation of the molecules which can be applied to the polar semiconducting molecules. Some of them are specific to organic molecular materials, but we hope that there are hints for the grain alignment of other materials.

ALIGNMENT BY SUBSTRATE SURFACES WITH NANOSCALE TEMPLATES

First, we will discuss the effect of the morphology of the substrate surface. There are reports on the alignment of organic crystals or polymers by modifying the surface morphology, i.e. rubbing of the surface for liquid crystal devices, friction transfer of polymers [1] and deposition of small molecules on it [2] and graphoepitaxy using electron-beam lithography of the template[3]. The degree of grain alignment by these techniques is enough for some applications but is not perfect. For example, Fig. 1 shows the reflection high energy electron diffraction (RHEED) of a pentacene thin film grown on friction-transferred poly-tetrafuluoroethylene film on Si substrate and polycrystalline Debye-Scherer rings are observed with spots and streaks from ordered grains. Since we are interested in the quantitative measurement of the band structure of organic semiconductors by angle resolved photoemission measurements[4], we have pursued higher degrees of crystallinity and alignment.

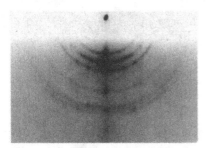

Figure 1. RHEED image of a pentacene film grown on a friction transferred polymer substrate.

Our method is to use well defined nanoscale template formed by step bunching of Si(111) polished with slight off angles[4]. The dangling bonds on the Si surface was terminated by 1/3 monolayer covered Bi adsorption[5]. By changing the annealing conditions in ultrahigh vacuum, it is possible to control the surface morphology. Regularly spaced steps with controlled height and terrace width can be fabricated on the surface. Figure 2 show the surface morphology of the pentacene films grown on the substrates prepared by different cooling conditions. It is easily understood that by changing the surface morphology of the substrates, the grain (thickness is a single molecular layer (1.5nm)) shape is greatly affected. The crystal alignment of the films was characterized by electron diffraction technique. It was revealed that the film shown in Fig. 2 (c) was almost single crystalline from the RHEED measurement [6]. The schematic alignment of crystal grains are illustrated in Fig. 3. The original alignment induced by the epitaxial relation with the substrate lattice was twelve-fold, but it was restricted to 3.5°- rotated twins. The threshold height of the surface steps for the alignment was ~1.5 nm, which correspond to the height of one molecule.

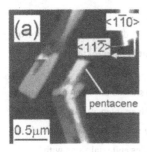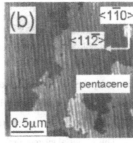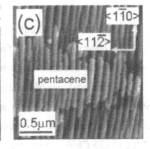

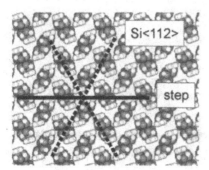

Figure 3. Schematic illustration of the relationship of pentacene lattice and the substrate lattice. The pentacene lattice can be rotated by multiples of 60° on flat substrate due to the symmetry mismatch (Si crystal axes direction are indicated by lines and 3.5° rotated twins are not shown). By introducing bunched steps on the substrate, the solid line marked by "step" restricts the orientation as shown in the figure.

The above result was obtained single crystalline substrate. It is still difficult to perfectly align the crystal orientation of organic thin films on amorphous substrates, but it has been revealed that atomic scale structure of substrate surface plays a very important role as well as morphology[7].

EFFECT OF ELECTRIC FIELD

It is easy to add electric dipoles to molecules. This feature is distinctively different from inorganic materials composed of atoms, and it opens an opportunity to control the grain/molecular orientation by external electric field.

As a model case of the alignment by the electric field, we studied the vacuum deposition of liquid crystal molecules. We applied a direct electric field with an intensity gradient during the thin film deposition. The used liquid crystal and the substrate was dodecyl-4-cyanobiphenyl (12CB) and KCl(001), respectively. This material system provides epitaxial films with a characteristic morphology dependent upon the growth temperature [8]. The shape of the molecule and the epitaxial relation on KCl(001) is shown in Figs. 4(a) and (b). The experimental set-up is shown in Fig. 4(c), in which inhomogeneous field was achieved by using needle and ring electrodes.

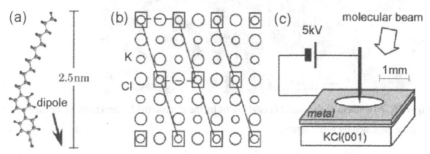

Figure 4. (a) Molecular structure of 12CB. (b) Epitaxial relation of 12CB on KCl(001). (c) Experimental setup.

The needle stem had a 1.5 mm diameter and the radius of the needle contact area with the surface was typically 100 μm, as measured using a scanning electron microscope. The radius of the hole in the plate electrode was 3 mm. The needle was connected to a DC voltage source (maximum 5 kV) during the deposition of the liquid crystal. The liquid crystal 12CB was introduced into the vacuum system using a load-lock mechanism and evaporated from a Knudsen cell at 50–53 °C. After deposition, the film morphology was examined using AFM. The intensity of the lateral electric field E and the field gradient dE/dr was estimated as a function of the distance from the edge of the needle electrode (r) using a finite element program. Figure 5 shows AFM images of films grown under an electric field with small coverage ($E = 4.4 \times 10^6$ Vm^{-1}, 22.5 °C). It is noted that needle-shaped islands were aligned with the direction of the electric field. This alignment was observed where E was high, while round-shaped islands were seen where E was low. The threshold field strength determined from the AFM ranged from 4×10^5 to 5×10^6 Vm^{-1} depending upon the temperature. The threshold value appears to increase as a linear function of temperature.

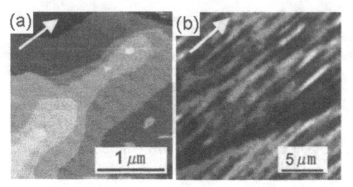

Figure 5. Alignment of the liquid crystal grains by the electric field. Directions of the electric field are indicated by white arrows.

In order to interpret the present results, the energy difference (Δ) between opposite alignment was estimated. It was $\sim 1.6 \times 10^{-10} E$ eV, and is smaller than the thermal energy ($k_B T = 3 \times 10^{-2}$ eV) by a factor of $40 \sim 500$. This indicates that the alignment of the islands is not a result of the alignment of a single molecular dipole but is the result of the concerted alignment of a number of dipoles ($40 \sim 500$). Therefore, the rheological properties of a liquid crystal must be taken into consideration in order to evaluate the temperature dependence of the threshold field strength for alignment. The linear dependence of the threshold field on the substrate temperature can be explained by considering the phenomenological Ginzburg–Landau type free energy, which is discussed in detail in our previous paper [9].

When the amount of deposited molecules increased, aggregation to the needle apex was observed (Fig,6) . It means that the molecules were forced to move this distance. A layer-by-layer aggregation is observed when the field gradient (dE/dr) was not less than 5.1×10^9 Vm^{-2} at 22.4 °C, while a uniformly covered surface with line contrast parallel to the substrate lattice and the round-shaped protrusion with a monolayer height was observed when $dE/dr = 3.9 \times 10^8$ Vm^{-2}.

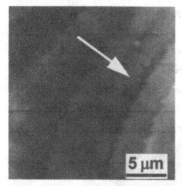

Figure 6. Layer by layer aggregation near the apex of the needle electrode.

Next, the force working on a 12CB molecule under the field gradient dE/dr was estimated, and the mechanism of layer-by-layer aggregation near the needle apex is considered. If the electric dipole of the migrating molecule is assumed to be aligned parallel to the electric field, the force F accelerating the molecule towards the direction of the higher electric field is given by $F=\mu\ dE/dr$. Under the given conditions, the threshold field gradient at 22.4 °C corresponds to 8.5×10^{-21} to 1.1×10^{-19} N. This is an extremely small value considering the drift velocity derived from the Einstein relationship and ordinary diffusion constant[9]. Molecular acceleration without friction for $10^{-5} \sim 10^{-4}$ s must be assumed to explain the result.

ALIGNMENT BY OPTICAL EXCITATION

Improvement of crystallinity by optical excitation during the growth has been reported on inorganic thin films and studied by various aspects. We have attempted to apply the optical excitation technique to tris-8-hydroxyquinoline-alminum (alq3) and found that the alignment of the crystal grains is controlled by the direction of the polarization of the incident light.   Alq3 is a Al-containing complex and has a permanent dipole.

It is known that alq3 also can be grown epitaxially as needle-shaped crystalline grains, but with multiple grain orientations (two with 90° rotation), on some alkali halide (001) substrates[10]. We attempted to optically excite the molecules by irradiating a continuous wave

(CW) He-Cd laser beam with 25 mW/cm^2. The substrate temperature was 80 °C to obtain well developed needle shaped grains.

Figure 7(a) shows the AFM image of the film grown without the laser irradiation. Since the [100] and [010] axes of the substrate are equivalent, needle shaped crystal grains are co-existent. Figure 7 (b) shows the AFM images with the laser beam whose polarization aligned with KCl[010]. Needle shaped grains are aligned parallel to the laser polarization. From the RHEED analysis, it was confirmed that the crystal orientation was aligned.

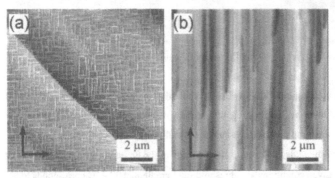

Figure 7. AFM images of alq3 films grown without(a) and with(b) laser irradiation. Black arrows show the [100] crystal axes of the substrate. The polarization of the laser in (b) is vertical direction in the figure.

The thickness dependence was examined to elucidate the mechanism of the laser orientation (Fig.8) The alignment ration defined by the areas of the islands aligned parallel to the laser polarization to the total area of the film was 75%, 91%, 100% at the thickness of 1nm, 4nm, and 50nm, respectively.

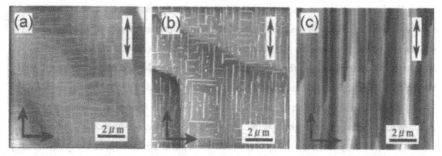

Figure 8. The thickness dependence of the alignment. (a) 1nm, (b) 4nm, (c) 50nm. The laser polarization is vertical direction in the figure and indicated by arrows.

When the laser polarization was rotated by 45°, or parallel to KCl[001], the morphology of the film was changed greatly. The alignment of the island did not occur, but the growth of the crystals is certainly affected as shown in Fig. 9.

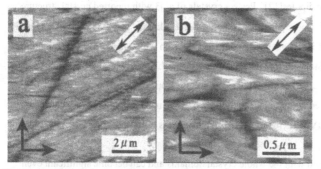

Figure 9. Effect of the laser irradiation polarized along KCl[110]. The arrows have the same meanings with Fig. 8.

The 442nm light is absorbed by the gas phase molecules and the solid alq3. When the film grown without the laser irradiation was irradiated by the same 442nm laser, no alignment was observed. Since the alignment did not occur by irradiation of 532 nm laser, the optical excitation during the deposition process seems critically important.

We here construct a model to explain all of the experimental result. As a first step, we estimate the ration of optically excited molecules migrating on the surface. The ratio is ~ 5 x 10-6, which is too small to consider dynamic effect of excitation dipoles that are seen in excimer or J-aggregate of dye molecules. The dynamic effect of the molecular excitation in the solid with photon frequency is also ruled out because the coherence length of the molecular excitation is 10-100 molecules and much shorter than the actually observed domain size. Thus the mechanism of the alignment is not dynamic but static effect induced by optical excitations.

The most easy but trivial effect is thermal. The optical excitation probability should have anisotropy and can heat the molecules or grains along the polarization. This mechanism is abandoned since it cannot explain the thickness dependence and the specific morphology observed with [110] excitation.

The model most plausible for us is related to the giant photovoltaic effect observed in ferroelectrics (to be more precise, pyroelectrics)[11]. There is no report that alq3 is ferroelectric, it is well established that vacuum deposited film easily acquire pyroelectricity, or spontaneous formation of macroscopic dipoles[12]. Perfect crystals of alq3 with reported crystal structure ($\alpha \sim \varepsilon$) do not have macroscopic permanent dipoles, but it is known that alq3 crystals can easily accommodate defects. It is reasonable that our epitaxial film with unknown crystal structure have permanent dipoles induced from defects. It is well established that polarized ferroelectrics shows giant photovoltaic effect comparable to the polarization when optically excited [13,14]. This effect can be observed with very weak light such as 0.2 mW/cm^2 and photovoltaic electric field as high as several MVm^{-1} is reported [14]. This effect is used as the basis of some types of photorefractive devices.

This effect is related with the whole crystal domains and can become significant even when the percentage of the optically excited molecules is extremely small as in this experiment. The electric field of MVm^{-1} is enough for the alignment of the surface migrating species as discussed in the previous section. This mechanism can explain the thickness dependence since the photo-induced electric field will become stronger with the thickness increase. It also can explain the effect of 45° -rotated polarization because electric field to the polarization will compete with the epitaxial relation.

Unfortunately, the alq3 films with perfect alignment fabricated in this experiment cannot be used as polarized electroluminescence devices because the anisotropy of the optical excitation is lost very quickly within the molecule[15]. However, when we regard the organic semiconductor molecule as a semiconductor nanocluster, it is attractive to imagine the possibility to easily add permanent dipoles by molecular designs.

The molecules have extra degree of freedom such as permanent dipoles and molecular shapes, and the artificial grain alignment technique not accessible to simpler materials can be applied. We have presented the effect of the nano-scale template as a kind of sophisticated graphoepitaxy, external electric field and optical excitation. Since the evaporated species in ionic crystals should have electric dipoles and the giant photovoltaic effect is observed in the ferroelectrics, some of these techniques will be applicable also to the inorganic materials.

The present work was partly supported by JST PRESTO and Seeds Innovation program, KAKENHI grants (17260008, 19651048) and Ogasawara foundation. Support from GCOE program (Chemistry Innovations, The University of Tokyo) is gratefully acknowledged.

1. J.C.Wittmann, P. Smith, Nature 352, 414 (1991).

2. H. Kihara, Y.Ueda, A, Unno, T. Hirai, Mol. Cryst. Liq. Cryst. 424, 195 (2004).

3. S. Ikeda, K. Saiki, K. Tsutsui, T. Edura, Y. Wada, H. Miyazoe, K. Terashima, K. Inaba, T. Mitsunaga, T. Shimada, Appl. Phys. Lett. 88, 251905 (2006).

4. T. Shimada, A. Suzuki, T. Sakurada, A. Koma, Appl. Phys. Lett. 68, 2502 (1996).

5. T. Shimada, M. Ohtomo, T. Suzuki, K. Ueno, S. Ikeda, K. Saiki, M. Sasaki, K. Inaba, Appl. Phys. Lett. in press.

6. T. Suzuki, T. Shimada, K.Ueno, S.Ikeda, K. Saiki and T. Hasegawa: Mater. Res. Soc. Conf. Proc. 965, S06-19 (2007).

7. S. Ikeda, K. Saiki, Y. Wada, K. Inaba, Y. Ito, H. Kikuchi, K. Terashima, T. Shimada, J. Appl. Phys. 103, 084313 (2008).

8. T.Shimada, M. Nagahori, A. Koma, Surf. Sci. 423, L285 (1999).

9. T.Shimada, M. Nagahori, A. Koma, Surf. Sci. 564, L263 (2004).

10. H. Ichikawa, T. Shimada and A. Koma: Jpn. J. Appl. Phys. 40, L 225 (2001).

11. H.Ichikawa, K.Saiki, T.Suzuki, T. Hasegawa and T. Shimada: Jpn. J. Appl. Phys. 44, L1469 (2005).

12. E. Ito, Y. Washizu, N. Hayashi, H. Ishii, N. Matui, K. Tsuboi, Y. Ouchi, Y. Harima, K. Yamashita and K. Seki: J. Appl. Phys. 92, 7306 (2002).

13. A.A.Grekov, M.A.Malitsjaya, V.D.Spitzina and V.M.Fridkin: Soviet Physics - Crystallography 15, 423 (1970).

14. P.S.Brody: Solid State Commun. 12, 673 (1973).

15. N. Ogawa , A. Miyata, H. Tamaru, T. Suzuki, T. Shimada, T. Hasegawa, K. Saiki and K. Miyano: Chem. Phys. Lett. 450, 335 (2008).

Mater. Res. Soc. Symp. Proc. Vol. 1150   1150-RR07-01

Inclined Substrate Deposition of Biaxially Textured Magnesium Oxide Films*

Beihai Ma[1], U. (Balu) Balachandran[1], Rachel E. Koritala[2], and Dean J. Miller[2]
[1]Energy Systems Division, [2]Materials Science Division,
Argonne National Laboratory, Argonne, IL 60439, U.S.A.

ABSTRACT

Inclined substrate deposition (ISD) has great potential for rapid production of high-quality textured template films. We have grown biaxially aligned magnesium oxide (MgO) films on metallic substrates by ISD at deposition rates, 20-100 Å/sec. Scanning electron microscopy of the ISD MgO films showed columnar grain structures with a roof-tile-shaped surface. X-ray diffraction and pole figure analysis revealed that the c-axis of the ISD MgO is tilted at an angle with respect to the substrate normal. Roughly a film thickness of 0.5 μm is required to obtain ISD MgO films with good biaxial texture. In-plane alignment, tilt angle, and surface roughness are dependents of inclination angle. With a 0.5 μm thick homoepitaxial coating over 1.5 μm thick ISD MgO, we measured full width at half maximum (FWHM) of ≈10° for samples grown with inclination angle >30°. Highly textured YBCO films were deposited by pulsed laser deposition (PLD) on the ISD MgO template films grown on metal tapes. The orientations of YBCO films are dependent on buffer materials. "c-axis aligned" YBCO films were obtained using yttria-stabilized zirconia (YSZ) and ceria (CeO_2) buffer layers. "c-axis tilted" YBCO films were fabricated on ISD MgO template using $SrRuO_3$ (SRO) and $SrTiO_3$ (STO) buffer films. Critical current density $J_c \approx 1.2 \times 10^6$ A/cm^2 and $J_c \approx 1.6 \times 10^6$ A/cm^2 were measured at 77 K in self-field on c-axis oriented and c-axis tilted YBCO films, respectively.

Keywords: ISD, MgO, texture, thin film, superconductor, oxide buffer

INTRODUCTION

The second-generation $YBa_2Cu_3O_{7-\delta}$ (YBCO)-coated conductor offers great promise for high-current carrying wires and other electric power devices operating at temperatures that approach liquid nitrogen [1-3]. Textured template films or buffer layers are needed for deposition of biaxially aligned YBCO films to overcome the weak links at high-angle grain boundaries and, therefore, to achieve high critical current density (J_c) in the YBCO films deposited on polycrystalline metallic substrates [4]. Several techniques, including ion-beam-assisted deposition (IBAD), rolling-assisted biaxially textured substrates (RABiTS), and inclined-substrate deposition (ISD), have been developed in recent years [5-9]. When compared with IBAD and RABiTS, the ISD process has many advantages. It produces biaxially textured films at high deposition rates (20-100 Å/sec) without the need of an assisting ion source and is independent of the recrystallization properties of the metallic substrates [9]. It is easy to scale up

*The submitted manuscript has been created by UChicago Argonne, LLC, Operator of Argonne National Laboratory ("Argonne"). Argonne, a U.S. Department of Energy Office of Science laboratory, is operated under Contract No. DE-AC02-06CH11357. The U.S. Government retains for itself, and others acting on its behalf, a paid-up nonexclusive, irrevocable worldwide license in said article to reproduce, prepare derivative works, distribute copies to the public, and perform publicly and display publicly, by or on behalf of the Government.

and capable of depositing biaxially textured films on curved surfaces [10]. Therefore, the ISD process is an attractive alternative for production of long-length coated conductor wires.

IBAD and RABiTS produce biaxially textured substrates with their c-axes normal to substrate surfaces. Thus, a cube-on-cube epitaxial growth of buffer layers and YBCO films will readily provide superconductors in which the c-axis is normally aligned. In the ISD process, the c-axis of the biaxially textured MgO is tilted from substrate normal; its terminal surface is rougher and has terraced structure. Therefore it is more challenging to fabricate high quality YBCO coated conductors on the ISD MgO buffered metal substrates.

We grew biaxially textured MgO thin films on polished Hastelloy C276 (HC) substrates by ISD with an electron beam (e-beam) evaporation system. Oxide buffers [yttria-stabilized zirconia (YSZ) with ceria (CeO$_2$) cap layer, strontium ruthenium oxide (SRO), strontium titanium oxide (STO)] and YBCO films were subsequently deposited on the ISD-MgO-buffered metallic substrates by pulsed laser deposition (PLD). The surface morphology was investigated by scanning electron microscopy (SEM). The surface roughness was measured by atomic force microscopy (AFM). X-ray pole figures, as well as ϕ- and ω-scans, were used to analyze texture. In this paper, we discuss the growth mechanism, microstructure, and dependence of biaxial alignment of ISD MgO films on film thickness and inclination angle. We also describe the orientation relationships and superconducting properties of YBCO deposited on HC substrates using two different ISD MgO architectures to produce "c-axis oriented" and "c-axis tilted" coated conductors.

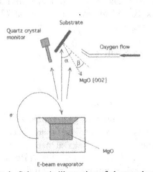

Figure 1. Schematic illustration of electron-beam evaporation system for ISD process.

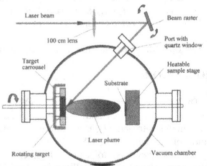

Figure 2. Schematic illustration of pulsed laser deposition system for deposition of oxide films.

EXPERIMENTAL PROCEDURE

A schematic illustration of the ISD setup is shown in Fig. 1. MgO thin films were grown from an MgO source by e-beam evaporation. Fused lumps of MgO (Alfa Aesar, 99.95% purity, 3-12 mm pieces) were used as the target material. Polished polycrystalline HC tapes were used as substrates. The substrates were mounted on a tiltable sample stage above the electron-beam evaporator. The substrate inclination angle α, defined as substrate normal with respect to the evaporation direction, was varied between 15 and 70° during our experiments. Oxygen flow was introduced into the system during film deposition. The base pressure of the vacuum chamber was ≈1 x 10^{-7} torr, which rose to 2 x 10^{-5} torr during deposition. A quartz crystal monitor was mounted beside the sample stage to monitor and control the deposition speed. High deposition rates of 20-100 Å/sec were used, and the substrate temperature was maintained between room temperature and 50°C during deposition. After the deposition of ISD films, a thin dense layer of MgO was deposited on top of the ISD film at a zero-degree inclination angle and an elevated deposition temperature (≈700°C). This dense layer has the same crystalline texture as the ISD MgO film, and is referred to as the homoepitaxial (HE) MgO layer.

A schematic illustration of the PLD system is shown in Fig. 2. Oxide buffer and YBCO films were deposited by a PLD system equipped with a Lambda Physik LPX210i excimer laser, with a Kr-F_2 gas premixture as the lasing medium. This device generates a pulsed laser beam of 248-nm wavelength and 25-ns pulse width at a repetitive rate up to 100 Hz. The oxide buffer and YBCO targets were 45 mm in diameter and 6 mm thick. The target carousel is capable of handling four different targets for deposition of multiple films without breaking vacuum. Substrates were attached to a heatable sample stage with silver paste and heated to a high temperature (700 - 800°C) during deposition. The size of the laser spot focused at the rotating target was ≈12 mm^2, which resulted in an energy density of ≈2.0 J/cm^2. The distance between the target and the substrates was ≈7 cm. The desired oxygen partial pressure was obtained by flowing ultra-high-purity oxygen through the chamber. Detailed experimental conditions for deposition YSZ/CeO_2, and $SrRuO_3$ (SRO), and $SrTiO_3$ (STO) buffers on ISD MgO template films were reported earlier [10-12].

The superconducting critical transition temperature (T_c) was determined by the inductive method, and J_c was measured by the standard four-probe transport method at 77 K in liquid nitrogen by the 1-μV/cm criterion over the entire sample width. Film texture was characterized by X-ray diffraction and pole figure analysis using Cu Kα radiation with a Bruker AXS D8 DISCOVER system equipped with GADDS. In-plane texture was characterized by the FWHM of φ-scans for the (002) reflection, and out-of-plane texture was characterized by the FWHM of ω-scans at the [001] pole for the same reflection. As for the YBCO films, in-plane texture was measured from the (103) φ-scan, and out-of-plane texture was measured from the (005) ω-scan. Surface roughness was measured by AFM using a Veeco Dimension D3100 Scanning Probe Microscope operated in tapping mode. SEM (Hitachi S-4700-II) was utilized to observe the surface and cross-sectional morphology. Crystalline orientation of textured films was determined from X-ray diffraction combined with pole figure analysis and was confirmed by transmission electron microscopy (TEM)/selected area diffraction (SAD) performed with a Philips CM30 electron microscope.

RESULTS AND DISCUSSION

Biaxially textured ISD MgO Template Films

Plan-view SEM (Fig. 3a) revealed a roof-tile structure for the ISD MgO film deposited at room temperature. Columnar grains were observed from the cross-sectional fracture surface

(Fig. 3b). The MgO grain width increased when the film grew for the first 0.5-μm thickness; it then became stabilized at ≈0.2 μm without noticeable change in size when the film grew further in thickness. Electron-beam evaporation of MgO at 700°C with a zero-degree inclination angle produced a smoother and denser homoepitaxial (HE) layer (shown in Figs. 3c and 3d). Figure 4 shows AFM surface images of the as deposited ISD MgO film (Fig. 4a) and the film after deposition of 0.5-μm-thick HE MgO at 700°C (Fig. 4b). Root-mean-square (RMS) surface roughness of ≈20 nm was measured on the as-deposited ISD MgO film by tapping-mode AFM. RMS surface roughness was reduced to ≈8 nm after the deposition of a HE MgO of 0.5-μm-thick at 700°C.

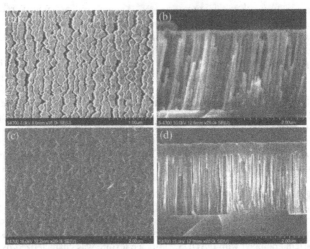

Figure 3. (a) Plan view and (b) cross-sectional SEM images of ISD MgO film deposited at room temperature with α = 55°; (c) plan view and (d) cross-sectional SEM images of MgO film after deposition of additional layer of MgO e-beam evaporated at 700°C with α = 0°.

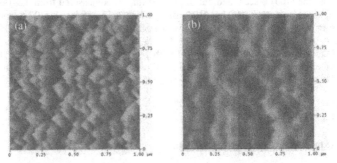

Figure 4. AFM surface images of (a) as deposited ISD MgO film and (b) after the deposition of 0.5-μm-thick homoepitexial MgO at 700°C.

A typical X-ray pole figure of the ISD MgO film deposited at room temperature with an inclination angle of 55° is shown in Fig. 5a. The ISD MgO film exhibits good texture; distinct in-plane alignment can be seen by the well-defined poles for not only the [001] axis but also the [010] and [100] axes. Unlike the YSZ films prepared by inclined-substrate PLD [13], where the (001) planes are nearly parallel to the substrate surface, the [001] axis of the ISD MgO buffer layer is tilted away from substrate normal. The asymmetric distribution of the pole peaks reveals that the MgO (001) plane is tilted. The tilt angle β can be measured from the angular distance between the strongest pole and the center on the MgO (002) pole figure as shown in Fig. 5a. We observed tilt angle β of ≈22° and ≈32° for samples deposited with inclination angle α of ≈35° and ≈55°, respectively. Figure 5b shows the TEM cross sectional image of an ISD MgO film along with selected area electron beam diffraction pattern taken on the grain. The terminal surface of ISD MgO film is indeed a (002) plane. The tilt angle can be measured from the degree of tilt for its c-axis from the substrate normal (SN), as shown in Fig. 5b.

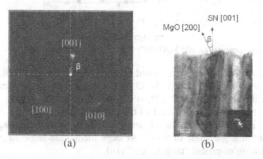

Figure 5. (a) MgO (002) pole figure and (b) TEM cross-sectional image and selected area diffraction pattern showing c-axis tilted nature for the ISD MgO films.

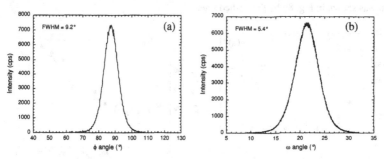

Figure 6. MgO (002) (a) φ-scan and (b) ω-scan patterns after homoepitaxial growth of 0.5-μm-thick MgO layer on ISD MgO film at elevated temperature.

Figure 6 shows the φ- and ω-scan patterns for MgO (002) after homoepitaxially growing a 0.5-μm-thick MgO layer on a ≈1.5-μm ISD MgO film at elevated temperature. This particular sample was deposited with an inclination angle of ≈55°. FWHMs of 9.2 and 5.4° were observed in the MgO (002) φ-scan and ω-scan, respectively. Figure 7 shows the in-plane texture (φ-scan FWHM) measured on ISD MgO films of various thicknesses. All samples were deposited with 55° inclination angle. The in-plane texture improves rapidly with increasing film thickness when the film thickness is below 0.5 μm thick, and then stabilized to a value of ≈9° with further increase in film thickness.

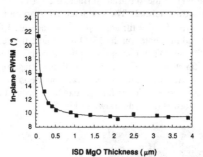

Figure 7. In-plane texture measured on ISD MgO (002) films of various thicknesses.

Figure 8 shows the in-plane texture and tilt angle measured on ISD MgO films deposited with various inclination angles. All ISD MgO films are 2 μm thick. In-plane texture improves with increasing inclination angle up to ≈55°, as shown in Fig. 8a. Within the experimental error, it appears there are two minima associated with inclination angle of 35 and 55°. Explanation is given in the following paragraph. The tilt angle decreases with decreasing inclination angle down to an inclination angle value of ≈30°. At large inclination angle (α ≥30°), experimental data obeys the following empirical "tangent rule" [14],

$$\tan(\beta) = \frac{1}{2} \cdot \tan(\alpha)$$

(1)

which was shown in Fig. 8b by the dashed line.

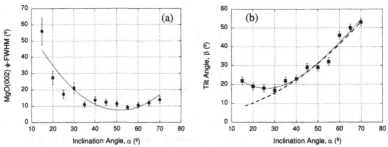

Figure 8. (a) In-plane texture and (b) tilt angle measured on ISD MgO (002) films deposited using various inclination angles.

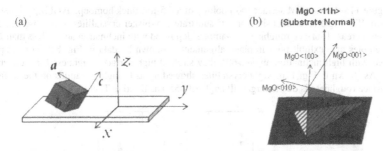

Figure 9. Illustration drawing of (a) MgO cube on the substrate and (b) relationship between ISD MgO axes and the substrate normal.

MgO has a sodium-chloride (NaCl) structure with a lattice constant of 4.203 Å. Its natural crystallites take cube shape. Grain texturing in ISD MgO films relies on shadowing and competitive growth. The "roof-tile" shaped ISD MgO film surface can be considered as a MgO cube grain rotated along its <110> axis; i.e., its <110> axis is parallel to the substrate plane and its <001> axis has a angle β (the tilt angle defined earlier) with respect to the substrate normal, as shown in Fig. 9a. MgO (001) plane turns towards MgO vapor flux to have a better acceptance to vapor molecules. From another point of view, the substrate normal is parallel to an MgO <11h> direction, as shown in Fig. 9b. Therefore the tilt angle is the angle between MgO <001> and MgO <11h>. Values of tilt angle for several primary (low index) planes are listed in Table 1 below. Comparing the experimentally measured tilt angle values with that listed in Table 1, we concluded that the ISD MgO films deposited with inclination angles of 35 and 55° are having MgO <112> and MgO <113> direction parallel to their substrate normal (SN) respectively. Those are the two lowest indexes listed in the table and yet not having three-fold symmetry. Films deposited with these two inclination angles (α = 35 and 55°) tend to have good biaxial alignment and, this is the reason why we observed local minima in Fig. 8a for two angles. We produced ISD MgO films with tilt angle β ≈55° (with <111> parallel to substrate normal) using an inclination angle α ≈70° as shown in Fig. 8b. Unfortunately, these films resulted 12-fold symmetry in the YBCO coated conductors and exhibited poor superconducting properties.

Table 1. Angles between <001> and <11h> in the cubic structure.

Plane index, 11h	Tilt angle β (°)
111	54.7
112	35.3
113	25.2
114	19.5
115	15.8

Figure 10 shows RMS surface roughness measured on ISD MgO samples deposited with various inclination angles. All samples have a 0.5-μm-thick homoepitaxial MgO layer on a ≈1.5-μm ISD MgO layer. Tapping mode AFM was performed over 2 μm x 2 μm areas. Samples with relatively smooth surface were prepared with inclination angles range from 20 to

60°. Figure 11a shows AFM surface morphology of a 0.5-μm-thick homoepitaxial MgO layer on a ≈1.5-μm ISD MgO layer deposited on HC substrates. Whisker crystallites were observed and lead to an increased surface roughness in samples deposited with inclination angles less than 20°. These samples also exhibit poor in-plane alignment as shown by data in Fig. 8a. For samples deposited with high inclination angle >60°, they showed high surface roughness for a different reason. As shown in Fig. 11b, MgO crystallites showed up as triangle pyramids on the surface. High surface roughness is due to larger tilt angle (β ≈55°) as listed in Table 1.

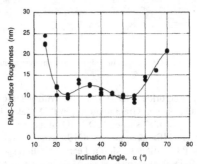

Figure 10. RMS surface roughness as a function of inclination angle determined by 2 μm x 2 μm AFM scans.

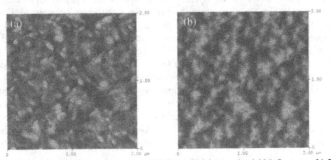

Figure 11. AFM surface morphologies of samples with 0.5-μm-thick homoepitexial MgO on top of 1.5-μm-thick ISD MgO deposited with inclination angle of (a) 15° and (b) 70°.

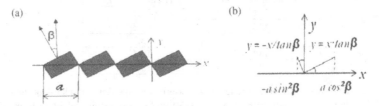

Figure 12. Illustrations of (a) one-dimensional terraced surface which resembles ISD MgO surface and (b) the coordinates of a unit cell used for roughness calculation.

RMS surface roughness is calculated by the following formula [15],

$$R_q = \sqrt{\frac{\sum_{i=1}^{n}(y_i - \bar{y})^2}{n-1}} \qquad (2)$$

where i refers to the i-th sampling point, y_i is the measured height at the i-th sample point, n the total number of samples, and $\bar{y}$ the average value of y_i. Equation 2 can also be written as,

$$R_q = \sqrt{\frac{\int (y - \bar{y})^2 dx}{\int dx}} \qquad (3)$$

where the integration is over the sampling area, or over the sampling length for one-dimensional case. When considering a simplified 1-d model illustrated in Fig. 12a and carrying out integration using Eq. 3 over a grain width (from $-a \cdot \sin^2\beta$ to $a \cdot \cos^2\beta$) as shown in Fig. 12b, we have RMS surface roughness proportional to $a \cdot \sin(2\beta)$ for samples with a columnar grain width of a and tilt angle β. This illustrated the fact higher tilt angle leads to a higher surface roughness for the ISD samples.

YBCO Deposited on ISD MgO with YSZ and CeO₂ Buffers

YSZ ($\approx$200 nm thickness) and CeO_2 ($\approx$10 nm thickness) films were epitaxially grown on top of the homoepitaxial MgO film by PLD at elevated temperatures (700-800°C). Both CeO_2 and YSZ layers have a cube-on-cube epitaxial relationship with the MgO film underneath. The (002) pole figures shown in Fig. 13 demonstrate a layer-by-layer epitaxy for MgO, YSZ, and CeO_2 films.

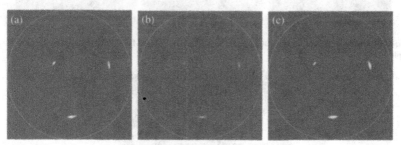

Figure 13. (002) pole figures for (a) MgO, (b)YSZ, and (c) CeO_2, showing cube-on-cube epitaxial relationship.

Figure 14 shows X-ray θ-2θ diffraction patterns for the YBCO films deposited on YSZ- and CeO_2-buffered ISD MgO substrates. The X-ray diffraction pattern measured with normal arrangement is shown in Fig. 14a. In this diffraction pattern, only YBCO (00l) peaks were observed; this finding indicates that the c-axis of YBCO films is parallel to substrate normal.

Note that the MgO, YSZ, and CeO$_2$ diffraction peaks cannot be seen in this pattern because their c-axes are tilted (at β ≈32°) from the substrate normal; therefore, the Bragg diffraction condition was not satisfied for these layers with the normal 2θ diffraction arrangement. (002) peaks for MgO, YSZ, and CeO$_2$ are observed from an X-ray θ-2θ diffraction taken at the strongest MgO (002) pole by tilting and rotating the sample to appropriate angles before measuring diffraction pattern, as shown in Fig. 14b. YBCO (00l) peaks are not visible in this diffraction pattern. This finding indicates that the c-axis of the YBCO film is not parallel to the c-axis of MgO, YSZ, or CeO$_2$.

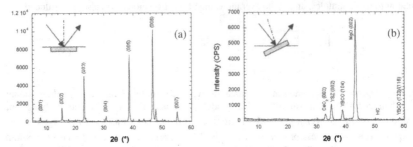

Figure 14. X-ray diffraction patterns at (a) normal arrangement and (b) MgO (002) pole.

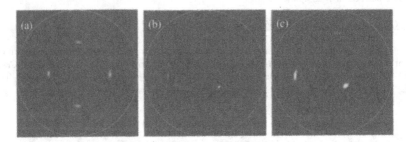

Figure 15. X-ray pole figures for (a) YBCO (113), (b) YSZ (002), and (c) MgO (002) measured on the same sample.

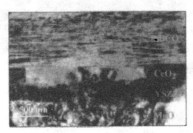

Figure 16. TEM cross-sectional image of YBCO film deposited on YSZ and CeO$_2$ buffered ISD MgO on HC substrate.

CeO$_2$ has a very close lattice match when compared with the YBCO pseudo cell, and the unit cell for the YBCO is larger than that for the CeO$_2$. Therefore, the X-ray diffraction peaks for CeO$_2$ are overlaid with YBCO peaks, and it is difficult to analyze the crystalline texture of the CeO$_2$ layer after deposition of YBCO films. Since the YSZ and CeO$_2$ layers were epitaxially grown on the MgO template before the deposition of the YBCO film, their crystalline texture should not have changed after the deposition. Hence, we tried to analyze the orientation relationship among the YBCO film, the YSZ buffer layer, and the MgO template layer. The YBCO (113) pole figure has four evenly distributed poles, as shown in Fig. 15a. This result indicates that the c-axis of YBCO is normal to the substrate surface. Furthermore, YBCO films were biaxially textured. The FWHM measured on YBCO (113) is $\approx$10°. The c-axes of YSZ and MgO are tilted, as shown in Figs. 15b and 15c. The pole figure for CeO$_2$ is not shown here because of its close lattice match with YBCO. The YBCO and MgO films were biaxially textured, as were the YSZ and CeO$_2$ films. We noted a unique orientation relationship among these films. From the relative positions of their poles, we derived the following orientation relationship: YBCO<100> // YSZ<111> // MgO<111>, and YBCO<010> // YSZ<110> // MgO<110>, or YBCO<001> // YSZ<112> // MgO<112>.

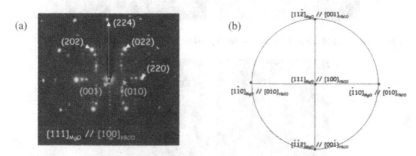

Figure 17. (a) TEM selected area diffraction along MgO [111] zone axis and (b) stereograph projection showing orientation relationship between YBCO and MgO films.

Figure 16 shows the TEM cross-sectional image of YBCO deposited on YSZ and CeO$_2$ buffered ISD MgO on HC substrate. As shown in figure, layer thickness for YSZ and CeO$_2$ are both $\approx$100 nm in the sample which was used for TEM selected area electron diffraction analysis. Selected area diffraction along the MgO and YSZ [111] zone axis is shown in Fig. 17a. A stereograph projection, shown in Fig. 17b, revealed orientation relationships between the YBCO and ISD MgO films. TEM results confirmed the orientation relationship between YBCO and MgO that was derived from X-ray pole figure analysis, with YBCO<100> // MgO<111> and YBCO<010> // MgO<110> or YBCO<001> // MgO<112>. With YSZ and CeO$_2$ buffer, c-axis oriented YBCO can be grown on c-axis tilted ISD MgO template, i.e., the YBCO c-axis is parallel to substrate normal rather than the c-axis of ISD MgO template film. YBCO coated conductors fabricated with such an architecture exhibited a sharp superconducting transition with T$_c$ = 91 K. Transport J$_c$ = 1.2 x 10^6 A/cm^2 at 77 K in self-field was measured on a short sample that was 0.4-μm thick, 4-mm wide, 1-cm long.

161

YBCO Deposited on ISD MgO with SrRuO₃ Buffers

Figure 18a shows the plan-view SEM image of SRO film that was epitaxially grown on top of the homoepitaxial MgO film by PLD at elevated temperatures (700-800°C) in 50 mTorr oxygen using a reprate of 4 Hz. SRO has an orthorhombic structure (JCPDS 89-5713) with lattice parameters a = 5.537 Å, b = 7.582 Å, and c = 5.573 Å. Its perovskite pseudo-cell parameter (a_p =3.92 Å) is in between that of YBCO and MgO. It is a conductive buffer material for epitaxial growth of YBCO on MgO template films. The top surface morphology of YBCO deposited on SRO-buffered ISD MgO is shown in Fig. 18b.

Figure 18. Plan view SEM images of (a) SRO buffer film and (b) YBCO on SRO buffered ISD MgO.

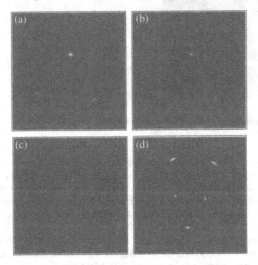

Figure 19. XRD pole figures measured for (a) MgO(002), (b) SRO(020), (c) YBCO(005), and (d) YBCO(103) showing layer-by-layer epitaxial relationship.

Figure 19 shows x-ray diffraction pole figures measured for MgO (002), SRO (020), YBCO (005), and YBCO (103). Similar to the MgO(002) pole figure shown in figure 19a, the

SRO (020) pole figure (shown in Fig. 19b) has three distinct poles, which are unevenly distributed; this finding indicates that the SRO film is biaxially textured and c-axis tilted. The c-axis of YBCO is also tilted (as shown by a single off-centered pole in Fig. 19c) and biaxially textured (as exhibited by the YBCO (103) pole figure shown in Fig. 19d). Pole figures shown in figure 19 demonstrated a layer-by-layer epitaxial relationship among MgO, SRO, and YBCO films. The c-axis of YBCO is tilted at the same β angle as that of the MgO template film deposited on the flat metallic substrate by ISD. Layer-by-layer ϕ-scans on samples of YBCO deposited on SRO-buffered ISD-MgO showed that the biaxial alignment improved with the deposition of each additional layer. A typical MgO (002) ϕ-scan FWHM is $\approx 10°$ for samples with $\alpha = 35°$, and FWHMs for SRO (020) and YBCO (005) are $\approx 7°$ and $\approx 6°$, respectively. In-plane texture of YBCO as determined from the ϕ-scan FWHM of YBCO (103) is $\approx 6°$. The biaxial alignment of the YBCO film is significantly improved over its MgO template by using the SRO buffer.

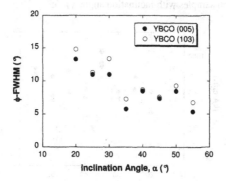

Figure 20. Biaxial alignment of YBCO grown on SRO buffered ISD-MgO as a function of inclination angle.

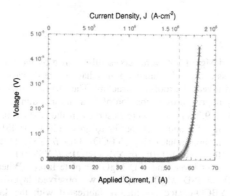

Figure 21. Current-voltage curve of YBCO on SRO buffered ISD MgO measured at 77 K.

The biaxial alignment determined from YBCO (005) and YBCO (103) φ-scans is plotted in Fig. 20 as a function of inclination angle used for deposition of ISD MgO films. In-plane alignment generally improves with increasing inclination angle. Two minima were observed at α = 35 and 55°, respectively. Using the four-probe DC method, we measured the highest transport J_c = 1.6 × 10^6 A/cm^2 and I_c > 110 A/cm at 77 K in self-field on sample with 35° inclination angle. Figure 21 shows the voltage versus current curve measured over the entire 0.5 cm width for the YBCO-coated conductor sample deposited on an SRO-buffered ISD-MgO (α = 35°) template grown on a flat Hastelloy C substrate. The thickness of this YBCO film was ≈0.68 μm. Figure 22 shows the transport J_c measured at 77 K in self field on YBCO superconductors deposited with SRO-buffered ISD-MgO templates grown on Hastelloy C substrates. The ISD-MgO template films were deposited with various inclination angles by electron beam evaporation. The SRO buffer layer and YBCO thin films were deposited by PLD. Maximum J_c values were measured on samples with inclination angle α = 35°.

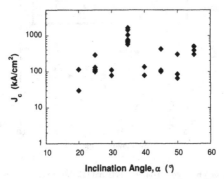

Figure 22. Transport J_c measured at 77 K on YBCO deposited on SRO buffered ISD-MgO of various inclination angles.

CONCLUSIONS

Biaxially textured MgO films were successfully grown by the ISD method. MgO films grown by the ISD process contained columnar grains that were terminated at the surface by (002) planes. Plan-view SEM revealed a roof-tile structure. The surface roughnesses of the ISD MgO films were significantly improved by deposition of an additional thin layer of MgO at elevated temperature. FWHMs of 9.2 and 5.4° were observed in the MgO (002) φ-scan and ω-scan, respectively. When using YSZ and CeO$_2$ buffer to grow YBCO on ISD MgO templates, we observed a unique orientation relationship, YBCO<100> // YSZ<111> // MgO<111> and YBCO<010> // YSZ<110> // MgO<110>, or YBCO<001> // YSZ<112> // MgO<112> among the YBCO film, YSZ buffer layer, and ISD MgO template layer. When using SRO or STO buffer to grow YBCO on ISD MgO templates, we observed layer-by-layer cube-on-cube epitaxial relationship. YBCO-coated conductors fabricated with the ISD MgO architecture exhibited T_c = 91 K with a sharp transition. Transport J_c of ≈1.2 x 10^6 A/cm^2 and ≈1.6 x 10^6

A/cm^2 were measured at 77 K in self-field on "c-axis oriented" YBCO-coated conductor (grown on the YSZ/CeO$_2$ buffered ISD MgO) and "c-axis tilted" YBCO-coated conductor (grown on SRO buffered ISD MgO) samples, respectively.

ACKNOWLEDGMENTS

SEM/TEM analysis was performed at the Electron Microscopy Center for Materials Research at Argonne National Laboratory. This work was supported by the U.S. Department of Energy under Contract DE-AC02-06CH11357.

REFERENCES

1. D. K. Finnemore, K. E. Gray, M. P. Maley, D. O. Welch, D. K. Christen, and D. M. Kroeger, "Coated Conductor Development: An Assessment," *Physica C*, 320, 1-8 (1999).
2. Y. Iijima and K. Matsumoto, "High-Temperature-Superconductor Coated Conductors: Technical Progress in Japan," *Supercond. Sci. Technol.*, 13, 68-81 (2000).
3. J. O. Willis, P. N. Arendt, S. R. Foltyn, Q. X. Jia, J. R. Groves, R. F. DePaula, P. C. Dowden, E. J. Peterson, T. G. Holesinger, J. Y. Coulter, M. Ma, M. P. Maley, and D. E. Peterson, "Advance in YBCO-Coated Conductor Technology," *Physica C*, 335, 73 (2000).
4. D. Dimos, P. Chaudhari, and J. Mannhart, "Superconducting Transport Properties of Grain Boundaries in YBa$_2$Cu$_3$O$_7$ Bicrystals," *Phys. Rev. B*, 41, 4038-4049 (1990).
5. Y. Iijima, K. Kakimoto, Y. Sutoh, S. Ajimura, and T. Saitoh, "Development of long Y-123 coated conductors by ion-beam-assisted-deposition and pulsed laser deposition method," *Supercond. Sci. Technol.*, 17, 264–8 (2004).
6. C. P. Wang, K. B. Do, M. R. Beasley, T. H. Geballe, and R. H. Hammond, "Deposition of In-plane Textured MgO on Amorphous Si3N4 Substrate by Ion-Beam-Assisted Deposition and Comparisons with Ion-Beam-Assisted Deposited Yttria-Stabilized-Zirconia," *Appl. Phys. Lett.*, 71, 2955-2958 (1997).
7. A. Goyal, D. P. Norton, J. D. Budai, M. Pranthaman, E. D. Specht, D. M. Kroeger, D. K. Christen, Q. He, B. Saffian, F. A. List, D. F. Lee, P. M. Martin, C. E. Klabunde, E. Hardtfield, and V. K. Sikka, "High Critical Current Density Superconducting Tapes by Epitaxial Deposition of YBCO Films on Biaxially Textured Metals," *Appl. Phys. Lett.*, 69, 1795-1797 (1996).
8. M. Bauer, R. Semerad, and H. Kinder, "YBCO Films on Metal Substrates with Biaxially Aligned MgO Buffer Layers," *IEEE Trans. Appl. Supercond.*, 9, 1502-1505 (1999).
9. B. Ma, M. Li, Y. A. Jee, B. L. Fisher, and U. Balachandran, "Inclined Substrate Deposition of Biaxially Textured Magnesium Oxide Films for YBCO Coated Conductors," *Physica C*, 366, 270-276 (2002).
10. B. Ma, M. Li, R. E. Koritala, A. R. Markowitz, R. A. Erck, R. Baurceanu, S. E. Dorris, D. J. Miller, and U. Balachandran, "Pulsed Laser Deposition of Biaxially Textured YBCO Films on ISD MgO Buffered Metal Tapes," *Supercond. Sci. Tech.*, 16, 464-472 (2003).
11. M. Li, B. Ma, R. E. Koritala, B. L. Fisher, X. Zhao, V. A. Maroni, S. E. Dorris, and U. Balachandran, "c-Axis orientation control of YBa$_2$Cu$_3$O$_{7-x}$ films grown on inclined-substrate-deposited MgO-buffered metallic substrates," *Solid State Communications*, 131, 101-105 (2004).

12. B. Ma, R. E. Koritala, B. L. Fisher, K. K. Uprety, R. Baurceanu, S. E. Dorris, D. J. Miller, P. Berghuis, K. E. Gray, and U. Balachandran, "High critical current density of YBCO coated conductors by inclined substrate deposition," *Physica C*, 403, 183-190 (2004).

13. K. Hasegawa, K. Fujino, H. Mukai, M. Konishi, K. Hayashi, K. Sato, S. Honjo, Y. Sato, H. Ishii, and Y. Iwata, "Biaxially Aligned YBCO Film Tapes Fabricated by All Pulsed Laser Deposition," *Appl. Supercond.*, 4, 487-493 (1996).

14. A. G. Dirks and H. J. Leamy, "Columnar microstructure in vapour-deposited thin films," Thin Solid Films 47, pp.219-233 (1977).

15. See for example, Military Standard MIL-STD-10A, "Surface Roughness, Waviness, and Lay," Department of Defense, U.S. Government printing office, Washington DC, 1963.

Mater. Res. Soc. Symp. Proc. Vol. 1150 © 2009 Materials Research Society   1150-RR04-04

Low-Temperature Selective Growth of Heteroepitaxial α-Al₂O₃ Thin Films on a NiO Layer by the Electron-Beam Assisted PLD Process

Makoto Hosaka[1], Yasuyuki Akita[1], Yuki Sugimoto[1], Yushi Kato[1], Yusaburo Ono[1], Akifumi Matsuda[1], Koji Koyama[2], and Mamoru Yoshimoto[1]

[1]Department of Innovative and Engineered Materials, Tokyo Institute of Technology, 4259-J2-46 Nagatsuta, Midori, Yokohama 226-8503, Japan

[2]Crystal Growth Laboratory, Namiki Precision Jewel Co., Ltd., 3-8-22 Shinden, Adachi, Tokyo 123-8511, Japan

ABSTRACT

Selective heteroepitaxial growth of α-Al₂O₃ thin films on a NiO layer was investigated using an electron-beam assisted pulsed laser deposition process. The epitaxial NiO layer was grown on an ultrasmooth sapphire (α-Al₂O₃ single crystal) (0001) substrate. The α-Al₂O₃ thin film could be grown epitaxially only in the electron-beam irradiated region of the epitaxial NiO layer at 300°C, while the amorphous Al₂O₃ film was grown in the non-irradiated region. The homoepitaxial growth of α-Al₂O₃ thin films could not be attained on the sapphire (0001) substrate at 300°C. This indicates that the electron-beam irradiation enhances heteroepitaxial growth of the α-Al₂O₃ thin films on the NiO layer at 300°C. When we annealed the epitaxial Al₂O₃/NiO bilayer film at 350°C in a hydrogen atmosphere, we could reduce only the NiO layer to an epitaxial Ni metal layer, allowing the fabrication of epitaxial Al₂O₃/Ni (insulator/metal structure) films. The fabricated Al₂O₃/Ni bilayer films exhibited a very smooth surface.

INTRODUCTION

Aluminum oxide (Al₂O₃) is widely used as a ceramic insulating material in electronic devices because it has a wide band gap (~7.0 eV), a high dielectric constant (~10), and thermally and chemically stable properties [1]. Recently, it has attracted considerable attention as insulating layers in tunnel magneto-resistance (TMR) devices [2–5], metal–oxide–semiconductor field-effect transistors (MOSFETs) [6], and other devices. For electronic applications, it is important to fabricate the heteroepitaxial layer structures such as multioxide and oxide/metal hybrid layers.

Using various deposition techniques, crystal or heteroepitaxial growth of Al₂O₃ films has been reported by several groups. The procedures mostly used for Al₂O₃ film growth are Al deposition in O₂ atmosphere at high temperatures [7] or oxidation of Al or Al alloy and annealing for crystallization [8–12]. However, these fabrication processes were performed at relatively high temperatures (over 500°C). For electronic applications, low-temperature fabrication techniques of heteroepitaxial insulating films are quite desirable to develop smooth surfaces and sharp interfaces.

Previously, we reported the room-temperature homoepitaxial growth of α-Al₂O₃ films on an atomically stepped sapphire substrate using the electron-beam assisted process [13]. This method allows us to fabricate homoepitaxial Al₂O₃ films exhibiting a very smooth surface. We also developed a novel metal epitaxy method by reduction of epitaxial metal oxide, which enabled us to apply the epitaxial fabrication techniques of metal oxide films to the construction of metal epitaxy [14]. In the present study, we investigated fabrication of heteroepitaxial α-Al₂O₃ films on NiO layers using the electron-beam assisted pulsed laser deposition (PLD) process at low temperatures. We then prepared Al_2O_3/Ni (insulator/metal structure) film by employing our previously developed metal epitaxy method using selective reduction of the epitaxial Al_2O_3/NiO layer.

EXPERIMENT

The oxide thin films were fabricated by the PLD method in an ultra-high-vacuum chamber equipped with high-energy electron diffraction (RHEED). The atomically stepped sapphire substrates (atomic steps 0.2 nm high, ultrasmooth terraces 50–100 nm wide) were first obtained by annealing commercial mirror-polished sapphire single crystal substrates at 1000°C for 3 h in air [15]. A pulsed KrF excimer laser (wave length of 248 nm, pulse duration of 20 ns, energy density of 3.0 J/cm^2, and repetition rate of 5 Hz) was focused on the sintered NiO target or single crystal α-Al_2O_3 target. The distance between the substrate and the target was fixed at 6.5 cm. The NiO film was deposited on the stepped sapphire substrate. The substrate temperature and O_2 gas pressure during the NiO film growth were fixed at 200°C and 1.0×10^{-5} Torr (back pressure of 1.0×10^{-8} Torr), respectively. The α-Al_2O_3 film was then deposited on the NiO layer by ablating the α-Al_2O_3 target. When the α-Al_2O_3 film was grown, a part of the substrate was continuously irradiated with an electron beam from the RHEED gun (acceleration voltage of 25 kV, dose rate of ~5 mA/cm^2, and beam diameter of 20 μm). The substrate temperature was kept at 300°C and the atmosphere was under a pressure of 1.0×10^{-5} Torr of O_2 gas during the Al_2O_3 film growth. The prepared Al_2O_3/NiO bilayer film was then annealed at 350°C under 760 Torr of H_2 gas flow for 2 h. The crystallographic property of the fabricated films was characterized by *ex situ* X-ray diffraction (XRD) and *in situ* RHEED. The surface morphologies were observed in air by an atomic force microscope (AFM).

RESULTS AND DISCUSSION

Heteroepitaxial growth of Al_2O_3 films on NiO

The 20-nm-thick NiO layer could be deposited epitaxially on the stepped sapphire (0001) substrate at 200°C. RHEED and XRD measurements confirmed the (111)-oriented epitaxial growth of the NiO film. Figure 1 shows the RHEED pattern of the NiO layer. The same streaky RHEED pattern was observed at every 60° beam rotation, indicating that the NiO layer was grown in the in-plane three-folded symmetry. The XRD measurement revealed that the NiO films grew epitaxially on the sapphire substrate with crystallographic relationship between NiO [111] // sapphire [0001] and NiO [1–21] // sapphire [11–20].

Figures 2(a) and (b) represent RHEED patterns of the α-Al_2O_3 film fabricated on the epitaxial NiO layer at the substrate temperature of 300°C for the non-irradiated and the electron-beam irradiated region, respectively. As shown in Fig. 2, the amorphous α-Al_2O_3 film grew in the non-irradiated region (a), while the α-Al_2O_3 film was grown epitaxially only in the electron-beam irradiated region (b). In contrast, the homoepitaxial growth of α-Al_2O_3 thin films could not be attained on the non-irradiated sapphire (0001) substrate at 300°C. This indicates that the electron-beam irradiation enhances the heteroepitaxial growth of the α-Al_2O_3 thin films at 300°C. Figure 2(c) shows a schematic image of this electron-beam induced epitaxy method.

Figure 1. RHEED pattern of the NiO layer fabricated on the ultrasmooth sapphire substrate.

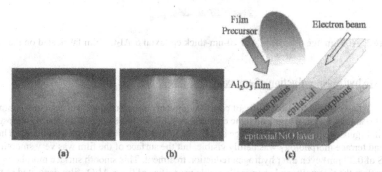

Figure 2. (a) RHEED pattern of the α-Al₂O₃ film deposited on the epitaxial NiO layer at 300°C for the non-irradiated region. (b) RHEED pattern for the electron-beam irradiated region under deposition. (c) Schematic image of the selective heteroepitaxial growth of the α-Al₂O₃ thin films by the electron-beam assisted PLD process.

It is well-known that the crystallization behavior of thin films depends strongly on the deposition temperature. The detailed mechanism of the present selective heteroepitaxial growth of α-Al₂O₃ by electron-beam irradiation is currently under investigation. The heteroepitaxial growth may be related to the localized heating effect as well as the defect formed by electron-beam irradiation, which enhances the structural change (from an amorphous to an epitaxial state) [16]. Thus, we can control crystallinity of the α-Al₂O₃ film and the crystallized area on the epitaxial NiO layer by applying electron-beam irradiation during the heteroepitaxial growth of the α-Al₂O₃ film.

Figure 3 shows AFM image of the 10-nm-thick epitaxial α-Al₂O₃ film on the NiO layer. A step and terrace morphology (step height of ~0.2–0.4 nm) inherited from the atomically stepped sapphire substrate was clearly observed on the surface. The flat surface of the epitaxial Al₂O₃ film may be due to the low-temperature growth at 300°C.

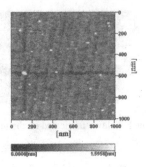

Figure 3. AFM image (1×1 μm^2) of 10-nm-thick epitaxial α-Al_2O_3 film fabricated on the NiO layer.

Selective hydrogen reduction of epitaxial Al_2O_3/NiO film

To investigate the possibility of constructing a heteroepitaxial bilayer structure composed of Al_2O_3 and Ni metal, we annealed the epitaxial Al_2O_3/NiO films at 350°C in the H_2 atmosphere for 2 h. Figure 4 represents AFM image of the Al_2O_3/NiO film after hydrogen reduction. The step and terrace morphology was hardly visible, but the surface of the film was very smooth (RMS of 0.30 nm) even after hydrogen reduction treatment. This smooth surface morphology may reflect the thermally and chemically stable properties of the α-Al_2O_3 film deposited as the upper layer.

Figure 4. AFM image (1×1 μm^2) of the Al_2O_3/NiO film annealed at 350°C in H_2 atmosphere.

Figure 5(a) shows XRD $2\theta/\theta$ profiles of the as-deposited epitaxial Al_2O_3/NiO films (bottom) and specimens after hydrogen reduction at 350°C (top). The diffraction peaks from NiO (111) and (222) planes disappeared after hydrogen reduction and the peaks from Ni (111) and (222) planes appeared for the hydrogen-reduced film (Figure 5(a) top). In contrast, peaks from Al metal were not observed. Figure 5(b) shows the RHEED pattern of the Al_2O_3/NiO film after hydrogen reduction. The streaky pattern close to that of the as-deposited epitaxial α-Al_2O_3 film indicated that the α-Al_2O_3 film was kept in the epitaxial state even after hydrogen annealing, as expected from the oxidation-reduction diagram (Ellingham diagram) of Al_2O_3/Al [17–20]. This selective hydrogen-reduction epitaxy method is considered to utilize the difference in reduction durability or stability of oxygen–metal bonding within the various oxides. Al ions have a higher adhesive energy to oxygen than Ni ions. Thus, in this case, only the NiO layer was reduced (to a Ni metal layer), while the Al_2O_3 layer remained in a metal oxide state.

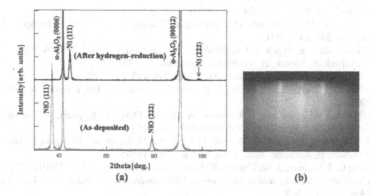

Figure 5. (a) XRD 2θ-θ profiles of the as-deposited epitaxial Al_2O_3/NiO film (bottom) and of the film annealed at 350°C in H_2 atmosphere (top). (b) RHEED pattern of the hydrogen-reduced Al_2O_3/NiO film.

CONCLUSIONS

Selective heteroepitaxial α-Al_2O_3 thin film was grown on the epitaxial NiO layer using an electron-beam assisted PLD process. We could grow the α-Al_2O_3 thin film epitaxially only in the electron-beam irradiated region of the epitaxial NiO layer at 300°C, while the amorphous Al_2O_3 film was grown in the non-irradiated region. When we annealed the epitaxial Al_2O_3/NiO bilayer film at 350°C in the hydrogen atmosphere, only the NiO layer was reduced, giving an epitaxial Ni metal layer, and a resulting epitaxial Al_2O_3/Ni (insulator/metal structure) film. The fabricated epitaxial Al_2O_3/Ni bilayer film exhibited a very smooth surface.

ACKNOWLEDGMENTS

This work was supported in part by the Ministry of Education, Culture, Sports, Science and Technology of Japan, the National Institute of Advanced Industrial Science and Technology of Japan, the New Energy and Industrial Technology Development Organization of Japan, and the Regional Innovation Creation R&D Program from the Ministry of Economy, Trade and Industry of Japan.

REFERENCES

1. K. Y. Gao, T. Seyller, L. Ley, F. Ciobanu, G. Pensl, A. Tadich, J. D. Riley and R. G. C. Leckey, Appl. Phys. Lett. **83**, 1830 (2003).

2. B. G. Park, T. Banerjee, J. C. Lodder, R. Jansen, Phys. Rev. Lett. **99**, 217206 (2007).
3. H. X. Wei, Q. H. Qin, M. Ma, R. Sharif, X. F. Han, J. Appl. Phys. **101**, 09B501 (2007).
4. X. F. Han, M. Oogane, H. Kubota, Y. Ando, T. Miyazaki, Appl. Phys. Lett. **77**, 283 (2000).
5. D. Wang, C. Nordman, J. M. Daughton, Z. Qian, J. Fink, IEEE Trans. Magn. **40**, 2269 (2004).
6. K. Yokoo, H. Tanaka, S. Sato, J. Murota, S. Ono, J. Vac. Sci. Technol. B, **11** (2), 429 (1993).
7. S. A. Chambers, Surf. Sci. Rep. **39**, 105 (2000).
8. R. Franchy, Surf. Sci. Rep. **38**, 195 (2000).
9. R. M. Jaeger, H. Kuhlenbeck, H.-J. Freund, M. Wuttig, W. Hoffmann, R. Franchy, H. Ibach, Surf. Sci. **259**, 235 (1991).
10. M. Klimenkov, S. Nepijko, H. Kuhlenbeck, H.-J. Freund, Surf. Sci. **385**, 66 (1997).
11. Y. Lykhach, V. Moroz, M. Yoshitake, Appl. Surf. Sci. **241**, 250 (2005).
12. T. T. Lay, M. Yoshitake, W. Song, Appl. Surf. Sci. **239**, 451 (2005).
13. A. Sasaki, H. Isa, J. Liu, S. Akiba, T. Hanada, M. Yoshimoto, Jpn. J. Appl. Phys. **41**, 6534 (2002).
14. A. Matsuda, M. Kasahara, T. Watanabe, W. Hara, S. Otaka, K. Koyama, M. Yoshimoto, Mater. Res. Soc. Symp. Proc. **962**, 0962-P09-04 (2007).
15. M. Yoshimoto, T. Maeda, T. Ohnishi, O. Ishiyama, M. Shinohara, M. Kubo, R. Miura, A. Miyamoto, H. Koinuma, Appl. Phys. Lett. **67**, 2615 (1995).
16. J. Liu, C. J. Barbero, J. W. Corbett, K. Rajan, H. Leary, J. Appl. Phys. **73**(10), 5272 (1993).
17. D.R. Gaskell, *Introduction to Metallurgical Thermodynamics* (McGraw-Hill Kogakusha, Tokyo, 1973) p. 269.
18. P.K. Datta, H.I. Du, J.S. Burnell-Gray, R.E. Ricker, in *ASM Handbook: Volume 13B: Corrosion: Materials*, edited by S.D. Cramer, Bernard S., J. Covino (Asm International, Ohio, 1995), p. 490-495.
19. M.M. Gasik, M.I. Gasik, in *Handbook of Aluminum: Volume 2: Alloy Production and Materials Manufacturing*, edited by G.E. Totten, D.S. MacKenzie (Marcel Dekker Inc., New York, 2003), p. 48-69.
20. G. Kim, Y. Moon, D. Lee, J. Power Sources **104**, 181 (2002).

Mater. Res. Soc. Symp. Proc. Vol. 1150 © 2009 Materials Research Society   1150-RR04-06

Intensity-Modulated Excimer Laser Annealing to Obtain (001) Surface-Oriented Poly-Si Films on Glass: Molecular Dynamics Study

Norie Matsubara, Tomohiko Ogata, Takanori Mitani, Shinji Munetoh, Teruaki Motooka
Department of Materials Science and Engineering, Kyushu University, 744 Motooka, Fukuoka 819-0395, Japan

ABSTRACT

We have investigated the dependence of the melting and crystal growth rates on the crystal orientations at solid/liquid (s/l) silicon (Si) interfaces by molecular dynamics (MD) simulations. It was found that there was no appreciable difference in the melting rates in all crystal orientations, though the growth rate at Si(001)/l-Si was the largest in all crystal orientations. We have also performed MD simulations of intensity-modulated excimer laser annealing (IMELA) of Si thin films. These results suggest that (001) surface-oriented Si can be obtained by IMELA owing to the largest growth rate at Si(001)/l-Si of all in the repetitions of crystallization and melting.

INTRODUCTION

In order to develop "system-on- display" [1], high-quality polycrystalline silicon (*poly*-Si) thin films are required. Excimer laser annealing (ELA) is one of the most common methods for *poly*-Si growth on glass. It is reported that poly-Si with a grain size of 30 μm can be obtained by using phase filter ELA [2], but the crystal orientation of each grain cannot be controlled resulting in a spatial fluctuation of the threshold voltage. The molecular dynamics (MD) simulation is a powerful tool for analyzing crystallization processes in an atomic scale. Recently, we have investigated atomistic processes of nucleation and crystallization in super cooling liquid Si (l-Si) to analyze excimer laser induced growth of Si thin films based on MD simulations using Tersoff potential [3]. In this paper, we have investigated the dependence of the melting and crystal growth rates on the crystal orientations at solid/liquid (s/l) interfaces by MD simulations. We have analyzed the effect of the dependence of the crystal growth rates on the crystal orientations in the repetitions of crystallization and melting in nano-second scale. We have also investigated a new method to obtain (001) surface-oriented *poly*-Si films by intensity-modulated excimer laser annealing (IMELA).

CALCULATION METHOD

MD simulations have been performed using the Tersoff potential [4,5] that can well reproduce the properties of a-Si, crystalline Si (c-Si), and l-Si. Although the Tersoff potential predicts the melting point (T_m) of Si to be approximately ~ 2580 K, which is well above the experimental value of 1685 K, this is not a serious problem since it is possible to make a rescaling between the Tersoff temperatures and the real temperatures [6]. Atomic movements were determined by solving Langevin equations

$$m_i\ddot{r}_i(t)=-m_i\gamma_i\dot{r}_i(t)+F_i(t)+R_i(t) \tag{1}$$

where m_i is the atomic mass, $r_i(t)$ is the position vector of the ith atom at time t, $F_i(t)$ is the interatomic force calculated using the Tersoff potential, and γ_i and $R_i(t)$ are the friction constant and random force, respectively. In the Langevin equation, the temperature of the system can be controlled by friction and random force. We employed the scheme developed by van Gunsteren and Berendsen for numerical integrations of the Langevin equation [7]. The time step for the numerical calculations and the friction constant were set to these described in our previous paper (2.0 fs and 5.0×10^{12} s^{-1}, respectively) [8].

MELTING AND CRYSTAL GROWTH RATES

The melting rates at temperatures above T_m, 2600 ~ 3200 K and crystal growth rates at temperatures below T_m, 2000 ~ 2500K at Si(001)/l-Si and Si(111)/l-Si interfaces were examined by MD simulations as described previously [8]. The MD cell sizes and numbers of atoms in the present calculations are summarized in table I. Periodic boundary conditions were set in the X- and Y-directions. In the Z-direction, the atoms with Z coordinates less than 5 Å in the bottom of the MD cell were fixed, and the top surface was free to move. The melting and growth rates at Si(001)/l-Si and Si(111)/l-Si estimated from the interface displacement velocities for various temperatures are shown in Fig. 1. Though the melting rates are almost the same at temperatures above T_m, the growth rate at the Si(001)/l-Si interface was found to be larger compared to the Si(111)/l-Si interface at temperatures below T_m.

Table I. MD cell size and the number of atoms that were set to calculate the melting and crystal growth rates at Si(001)/l-Si and Si(111)/l-Si at various temperatures.

		MD cell size (Å)³	Number of atoms
Melting	Si(001)/l-Si	54.3 × 48.84 × 152.04	20160
	Si(111)/l-Si	53.21 × 46.08 × 150.50	18432
Crystal growth	Si(001)/l-Si	59.73 × 21.72 × 195.48	12672
	Si(111)/l-Si	59.86 × 26.88 × 197.53	15876

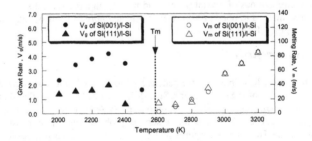

Figure 1. Melting and crystal growth rates of Si(001)/l-Si interface are circles, and Si(111)/l-Si interface are triangles.

The melting rates at 2700 K, and the growth rates at 2300K for Si (110), (112), (113), (115) l-Si interfaces were also examined by MD simulations. The variations in the melting rates were significantly smaller compared to variations in the crystal growth rates. The crystal growth rate at the Si(001)/l-Si interface was largest at 2300 K compared to the other crystal orientations (see table II).

There results suggest that the crystal growth rates depend on the crystal orientations and the growth rate at Si(001)/l-Si is largest in all orientations directions at any temperatures below T_m though the melting rates have no dependence on the crystal orientations at any temperatures above T_m.

Table II. Melting rates, crystal growth rates, and the numerical value of $(t_m/t_g)^{steady}$ at various s/l Interfaces.

	Si(001)/l-Si	Si(110)/l-Si	Si(111)/l-Si	Si(112)/l-Si	Si(113)/l-Si	Si(115)/l-Si
Melting rate at 2700 K (m/s)	6.12	6.63	5.12	5.31	5.50	6.03
Crystal growth rate at 2300 K (m/s)	4.19	2.64	2.03	2.19	2.56	3.36
$(t_m/t_g)^{steady}$	0.68	0.40	0.40	0.41	0.47	0.56

Intensity-modulated excimer laser annealing (IMELA)

Figure 2 shows the conditions of the temperature ($\propto$ the laser intensity) modulation during IMELA processes. T_{growth} is the temperature below T_m, $T_{melting}$ is the temperature above T_m, t_g is the holding time at T_{growth}, t_m is the holding time at $T_{melting}$, $GR(T_{growth})$ is the growth rate at T_{growth}, $MR(T_{melting})$ is the melting rate at T_m, respectively. In the simulations described later, T_{growth}, $T_{melting}$ and the frequency are assumed to be 2300K, 2700K and 1GHz, respectively.

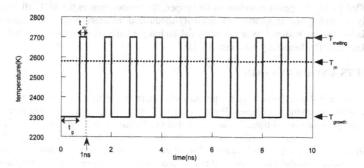

Figure 2. The set temperature change in the MD cell during IMELA processes.

The optimum conditions of the time ratio (t_m/t_g) in this assumption are derived. In the steady state (this means here that the nucleus repeats to crystal growth and melting by same length as centering on a certain standard position.), the following equation consists,

$$GR(T_{growth}) \times t_g^{steady} = MR(T_{melting}) \times t_m^{steady} \qquad (2)$$

where T_{growth} is the temperature below T_m, $T_{melting}$ is the temperature above T_m, t_g is the holding time at T_{growth}, t_m is the holding time at $T_{melting}$, $GR(T_{growth})$ is the growth rate at T_{growth}, $MR(T_{melting})$ is the melting rate at T_m, respectively.

The condition that crystalline nuclei at the Si(001)/l-Si interface prefer to grow whereas crystalline nuclei at the Si(111)/l-Si interface tend to melt during the repetitions of crystallization and melting is derived in the following expressions.

$$(t_m/t_g)^{steady}(hkl) = GR(T_{growth})/MR(T_{melting})(hkl) \qquad (3)$$

The numerical value of (t_m/t_g) can be adjusted up or down by varying the irradiated intensity-modulated excimer laser. If (t_m/t_g) is larger than (t_m/t_g)^{steady}, melting is prior to crystallization, contrary to this, if (t_m/t_g) is smaller than (t_m/t_g)^{steady}, crystallization is prior to melting. Only nucleus with the Si(001)/l-Si interface should be able to grow in the assumption of the laser within the range of a certain the time ratio (t_m/t_g), $0.56 < (t_m/t_g) < 0.68$ (see Table II).

We have performed MD simulations by IMELA of Si(001)/l-Si and Si(111)/l-Si thin films and analyzed the atomic processes. In regard to the IMELA of Si(001)/l-Si and Si(111)/l-Si thin films, the optimum condition of (t_m/t_g) was found to be from 0.40 to 0.68 (see table II), so that (t_m/t_g) was set as 0.4. The MD cell sizes and numbers of atoms in the present calculations are summarized in Table I. The initial MD cell contains c-Si/l-Si interface at about half of the MD cell. The initial MD cell is assumed that crystalline nuclei occur in l-Si (a-Si) during excimer laser irradiation. Periodic boundary conditions were employed in the X- and Y-directions. In the Z-direction, the atoms with Z coordinates less than 5 Å were fixed and the other atoms were controlled by the Langevin equation. In this paper, the temperature in the MD cell is controlled directly assuming the irradiation of intensity-modulated excimer laser. The laser intensity modulation can be realized by a conventional technique such as that using an electro optical modulator [9]. The frequency was set at 1 GHz.

RESULTS AND DISCUSSION

Figure 3 shows the average temperature change in all the atoms in the MD cell calculated by averaging their kinetic energies. Figure 4 (a) ~ (d) show snapshots of atomic configurations after annealing for typical time. It was found that crystalline nuclei at the Si(001)/l-Si interface prefer to grow, whereas crystalline nuclei at the Si(111)/l-Si interface tend to melt during the repetitions of crystallization and melting. The s/l interface moved up and down corresponding to the temperature modulation during IMELA processes, as shown Fig. 5. Because of the dependence of the crystal growth rate on the crystal orientations at s/l Si interface, it is suggested that (001)-oriented Si surface can be obtained by temperature modulation around the T_m while controlling the time ratio (t_m/t_g).

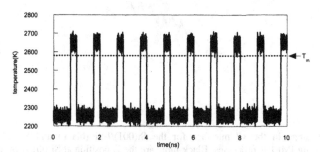

Figure 3. The average temperature change in all the atoms in the MD cell calculated by averaging their kinetic energies during IMELA processes.

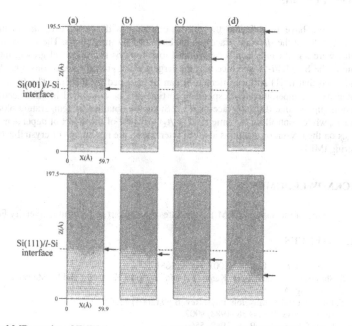

Figure 4. Typical MD results of IMELA processes at Si(001)/*l*-Si and Si(111)/*l*-Si interfaces: (a) before, and after annealing for (b) 3.8 ns, (c) 4.0ns, and (d) 9.8 ns. Crystal growth is dominant as the Si(001)/*l*-Si interface, while melting is rather prominent at Si(111)/*l*-Si interface as shown by the arrows indicating the *s/l* interface movement. The dotted lines are the initial interface positions.

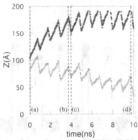

Figure 5. Change in the *s/l* interface for the Si(001)/*l*-Si (black) and Si(111)/*l*-Si (gray) interfaces during IMELA processes. Black plots are the Z-position at Si(001)/*l*-Si interface, and gray plots are the Z-position at Si(111)/*l*-Si interface. The time corresponding to the Fig. 4 (a) ~ (d) are shown with dashed lines.

CONCLUSIONS

We have investigated the crystal orientations dependence of the melting and crystal growth rates at the *s/l* Si interfaces on the basis of MD simulations. The variations in the melting rates were significantly smaller compared to variations in the crystal growth rates. The growth rate at the Si(001)/*l*-Si interface was the largest in all crystal orientations. Our MD simulations indicated that (001) surface-oriented Si can be obtained by IMELA. The laser energy ($\propto$ laser intensity $\propto$ temperature) corresponding to time can be controlled in nano-second scale. These results suggest that (001) surface-oriented Si can be obtained by temperature modulation around the T_m while controlling the time ratio (t_m/t_g), because of the effect of dependence of the growth rates on the crystal orientations at *s/l* Si interface in the repetitions of crystallization and melting during IMELA.

ACKNOWLEDGMENTS

One of the authors (N.M.) appreciates the support of Kyushu University Foundation.

REFERENCES

1. S. Uchikoga: MRS Bull. 27 (2002) No. 11, 881.
2. S. Shimoto, Y. Taniguchi, T. Katou, T. Endo, T. Ohno, K. Azuma and M. Matsumura: MRS. 2008 Spring Meeting. A12. 1
3. T. Motooka and S. Munetoh: Phys. Rev. B 69 (2004) 73307.
4. J. Tersoff: Phys. Rev. B 38 (1988) 9902.
5. J. Tersoff: Phys. Rev. B 39 (1989) 5566.
6. L. J. Porter, S. Yip, M. Yamaguchi, H. Kaburaki, and M. Tang: J. Appl. Phys. 81(1997) 96.
7. W. F. van Gunsteren and H. J. C. Berendsen: Mol. Phys. 45 (1982) 637.
8. S. Munetoh, T. Kuranaga, B. M. Lee, T. Motooka, T. Endo, and T.Warabisako: Jpn.J. Appl. Phys. 45 (2006) 4344.
9. A. Yariv and P. Yeh: *Optical Waves in Crystals: Propagation and Control of Laser Radiation (Wiley Series in Pure and Applied Optics)* (Wiley, New York, 2002) Chap. 9, pp. 318 and 366.

Mater. Res. Soc. Symp. Proc. Vol. 1150 © 2009 Materials Research Society   1150-RR04-07

Crystallization processes of amorphous Si during excimer laser annealing in complete-melting and near-complete-melting conditions: A molecular dynamics study

Tomohiko Ogata, Takanori Mitani, Shinji Munetoh, and Teruaki Motooka
Department of Materials Science and Engineering, Kyushu University, 744 Motooka, Fukuoka, 819-0395, Japan

ABSTRACT

We investigated crystallization processes of amorphous Si (a-Si) during the excimer laser annealing in the complete-melting and near-complete-melting conditions by using molecular dynamics simulations. The initial a-Si configuration was prepared by quenching liquid Si (l-Si) in a MD cell with a size of $50 \times 50 \times 150 \text{Å}^3$ composed of 18666 atoms. KrF excimer laser (wavelength: 248nm) annealing processes of a-Si were calculated by taking account of the change in the optical constant upon melting during a laser pulse shot with the intensity $I_o \exp[-(t-t_0)^2/\sigma^2]$ (I_o: laser fluence, t: irradiation time). The refractive indices of a-Si and l-Si were set at n+ik=1.0+3.0i and n+ik=1.8+3.0i, respectively. The simulated results well reproduced the observed melting rate and the near-complete-melting and complete-melting conditions were obtained for $I_o = 160\text{mJ/cm}^2$ and 180mJ/cm^2, respectively. It was found that larger grains were obtained in the near-complete-melting condition. Our MD simulations also suggest that nucleation occurs first in a-Si and subsequent crystallization proceeds toward l-Si in the near-complete-melting case.

INTRODUCTION

High-quality polycrystalline Si (*poly*-Si) is required to realize "System-in-display". Excimer laser annealing is an effective technique to obtain high-quality polycrystalline silicon (*poly*-Si) from amorphous silicon (a-Si) thin film deposited on glass or plastic substrates. Hatano *et al.* previously reported the relationship between grain sizes and excimer laser fluences [1]. In the lower fluence region, the grain size increases with increasing laser fluence. This irradiation condition gives rise to "partial-melting" of a-Si. On the other hand the grain size rapidly decreases with increasing laser fluence in the higher fluence region. This condition gives rise to "complete-melting" of a-Si. The transition region, which gives rise to "near-complete-melting" of a-Si, exists between "partial-melting" condition and "complete-melting" condition. We previously reported the mechanism of nucleation in the various laser fluences by MD simulations [2, 3]. Under the near-complete-melting condition, nucleation occurred predominantly in the unmelted a-Si region during the laser irradiation and then the crystal growth proceeds toward liquid Si region. In this paper, we performed the MD simulations of the nucleation and crystal growth processes during excimer laser annealing by taking account of reflection and absorption of a-Si and l-Si layers upon melting. We visualized the nucleation and crystal growth processes under each melting condition and discussed why the large grain can be obtained under the near-complete-melting condition.

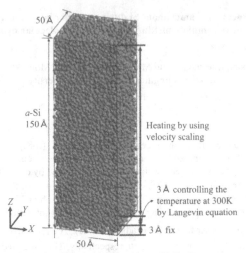

50 Å

a-Si
150 Å

Heating by using
velocity scaling

3 Å controlling the
temperature at 300K
by Langevin equation

Z
Y
X

3 Å fix

50 Å

Fig. 1 Initial MD cell obtained by rapid quenching of *l*-Si. Gray spheres indicate Si atoms.

CALCULATION METHOD

The initial *a*-Si configuration was prepared by quenching liquid Si (*l*-Si) in a MD cell with a size of $50 \times 50 \times 150 \text{Å}^3$ composed of 18666 atoms as illustrated in Fig.1. MD simulation was performed by using Tersoff potential [4, 5]. Although by using Tersoff potential the melting point (T_m) of crystalline Si is approximately 2580K, different to the experimental value of 1685K, the structures of *l*-Si or *a*-Si can be well reproduced with this potential. Periodic boundary conditions were used in X and Y directions. In the Z direction, atoms with the Z-coordinates less than 3 Å were fixed, while the top surface atoms of the *a*-Si layer were set to be free. The entire MD cell was heated at 300 K for 50 ps in order to relax structure with the time step of 2 fs. After structure relaxation of *a*-Si, we performed subsequent MD simulation in order to reproduce melting and crystallization under the excimer laser irradiation. KrF excimer laser (wavelength: 248nm) annealing of *a*-Si was estimated by using a Gaussian-shape laser pulse with FWHM of 25ns illustrated in Fig. 2. Figure 3 shows two layers model in order to estimate supplied laser energy. In the calculation, distinction between *a*-Si layer and *l*-Si layer was examined by diffusion coefficient of $8 \times 10^{-7} \text{ cm}^2 \text{s}^{-1}$. The refractive indices of *a*-Si and *l*-Si were set at n+ik=1.0+3.0i and n+ik=1.8+3.0i, respectively. The absorption of Si was set at $1.5 \times 10^6 \text{ cm}^{-1}$. The supplied laser energy was estimated by the two layers model and the region with Z-coordinates from 3 to 6 Å was controlled to 300 K by the Langevin equation with the friction constant γ set at $2.6 \times 10^{10} \text{ s}^{-1}$.

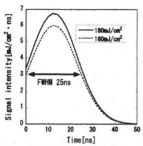

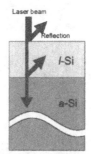

Fig.2 Intensity change of the laser pulse as a function of time. The shape is assumed to be the Gaussian function with the FWHM of 25 ns.

Fig.3 Schematic diagram of the two-layer model to calculate the deposited laser energy. Absorption in a-Si and l-Si layers as well as the reflection at the surface and interface is taken accounts.

RESULTS AND DISCUSSION

Figure 4 shows melting depth profile with the values of laser fluence, which are 160mJ/cm² and 180mJ/cm². The melting rate in the both cases is about 3m/s, which values are in good agreement with the experimental data about melting rate during irradiating excimer laser [1]. The case of laser fluence 180mJ/cm² was complete-melting condition because the whole a-Si was melted by the excimer laser pulse. On the other hand, the case of laser fluence 160 mJ/cm² was near-complete-melting condition because both l-Si and a-Si phases existed.

The large grains were obtained under the near-complete-melting condition while the spontaneous small grains in a-Si region were obtained under the complete-melting condition. Figure 5 shows diffusion coefficient and temperature profiles at 15 ns with excimer laser irradiated of fluence 160 mJ/cm². Above the Z=60 Å temperature is lower than melting point of Tersoff potential while diffusion coefficient is higher than the value of bulk l-Si. Therefore the phase of upper region is explained by the supercooled liquid. Growth rate of liquid phase epitaxy(LPE) is higher than that of solid phase epitaxy [2, 6].

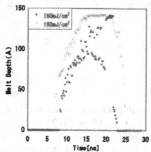

Fig.4 Change in the melt depth as a function of laser irradiation time.

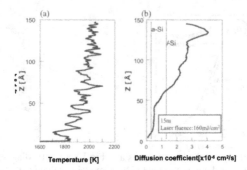

Fig.5 Temperature (a) and diffusion coefficient (b) profiles along the Z-coordinates at 15 ns for the laser of fluence 160 mJ/cm². The dotted lines in (b) represent the diffusion coefficient of bulk a-Si and l-Si obtained by our simulations.

Figure 6 shows the snapshots of atomic configuration during excimer laser irradiation with fluence 180 mJ/cm². The a-Si was melted from the upper surface region by the laser irradiation as shown in Fig.6 (b) and (c). The l-Si region was subsequently quenched and transformed to a-Si. After quenching the a-Si, spontaneous nucleation was occurred in the a-Si as shown in Fig.6 (e) and (f). Figure 7 shows the snapshots of atomic configuration during excimer laser irradiation with fluence 160 mJ/cm². The spontaneous nucleation was occurred in unmelted a-Si region during melting a-Si from upper region as shown in Fig.7(c). Figure 7(d) shows that liquid phase reached the nuclei and crystal growth was occurred. After that, explosive crystal growth toward l-Si phase was occurred as shown in Fig.7 (e) and (f). These results indicate that the nucleation in unmelted a-Si during laser irradiation and crystal growth toward supercooled l-Si are the reason of large grain in the near-complete-melting condition.

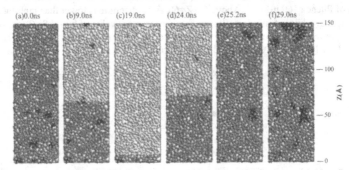

Fig.6 Snapshots of atomic configurations during the laser irradiation with the fluence 180 mJ/cm² : (a) initial, after (b) 9 ns, (c) 19 ns, (d) 24 ns, (e) 25.2 ns and (f) 29 ns. Amorphous Si atoms are shown in dark gray, liquid Si atoms are shown in light gray and well-ordered atoms are shown in black. These snapshots were obtained by projecting all the atoms in the MD cells in the 5 Å thick region from 45 to 50 Å in the X direction.

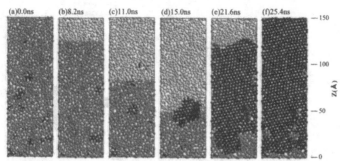

Fig.7 Snapshots of atomic configurations during the laser irradiation with the fluence 160 mJ/cm^2: (a) initial, after (b) 8.2 ns, (c) 11ns, (d) 15ns, (e) 21.6ns and (f) 25.4ns. Amorphous Si atoms are shown in dark gray, liquid Si atoms are shown in light gray and well-ordered atoms are shown in black. These snapshots were obtained by projecting all the atoms in the MD cells in the 5 Å thick region from 45 to 50 Å in the X direction.

CONCLUSIONS

We have performed MD simulations of excimer laser annealing of a-Si thin films by taking account of reflection and absorption. In the complete-melting condition, first a-Si was melted by laser irradiation and subsequently quenching l-Si was transformed to a-Si. Nucleation was occurred in the quenched a-Si and grain size is small in complete-melting condition. On the other hand in near-complete-melting condition, nucleation was occurred in the unmelted a-Si region during laser irradiation. The grain size was large in the near-complete-melting condition because crystal growth toward supercooled liquid Si.

REFERENCES

1. M. Hatano, S. Moon, M. Lee, K. Suzuki and C. P. Grigoropoulos: J. Appl. Phys. 87, 36 (2000).
2. S. Munetoh, T. Kuranaga, B. M. Lee, T. Motooka, T. Endo and T. Warabisako: Dig. Tech. Papers AM-LCD'05, 295 (2005).
3. S. Munetoh, T. Mitani, T.Kuranaga and T. Motooka: MRS Proc. 0958-L07-07(2006)
4. J. Tersoff: Phys. Rev. B 38, 9902 (1988).
5. J. Tersoff: Phys. Rev. B 39, 5566 (1989).
6. T. Motooka, K. Nisihira, S. Munetoh, K. Moriguchi and A. Shintani , Phys. Rev. B 61, 8537 (2000)

Mater. Res. Soc. Symp. Proc. Vol. 1150 © 2009 Materials Research Society   1150-RR04-16

Carbon Nanotube-Induced Changes of Crystal Growth In Polymer Films

Georgi Georgiev[1,2], Yaniel Cabrera[2], Lauren Wielgus[2], Zarnab Iftikhar[1], Michael Mattera[1], Peter Gati[1], Austin Potter[1] and Peggy Cebe[2]

[1]Department of Natural Sciences, Assumption College, Worcester, MA 01609
[2]Department of Physics and Astronomy, Tufts University, Medford, MA 02155

ABSTRACT

Isotactic Polypropylene (iPP) nanocomposites with low concentrations of multiwall carbon nanotubes (CNTs) 0-1% were studied, using differential scanning calorimetry and Avrami analysis. The nanocomposites were isothermally crystallized at 135°C, in order to measure the effect of nanotubes on the kinetics of crystallization. In our study there is a great effect of the CNTs on the iPP crystallization kinetics. The Avrami analysis shows increase in the crystallization rate constant and constancy the Avrami exponent with increase of the CNTs concentration. The full width at half maximum (FWHM) of the heat flow exotherm and the peak time for crystallization (t_p) change dramatically. For iPP, the carbon nanotubes serve as nucleation agents to speed up the crystallization process.

INTRODUCTION

Crystal nucleation processes and rates of growth in polymers are modified by the presence of carbon nanotubes (CNTs) [1]. We are exploring the effects of different concentrations of multiwall carbon nanotubes on the crystallization kinetics of isotactic polypropylene (iPP) isothermally crystallized at 135°C. The largest effect on kinetics that we observe is for concentrations of CNTs at or below 1% by mass, which is different than what other groups observe [2,3,4]. With increase of the concentration of nanotubes, we observe a decrease of the crystallization time during isothermal annealing, and an increase of the crystallization temperature during the nonisothermal crystallization. We analyze the nucleation and crystallization kinetics by using Kolmogorov-Johnson-Mehl-Avrami (JMAK) theory [5]. The Avrami exponent is not significantly affected by the addition of CNTs, but the rate constant K, the FWHM of the heat flow exotherm, the t_p and the crystallization half-time $\tau_{1/2}$ are affected greatly, results consistent with other reports [6].

THEORY

The theory of phase transitions is extremely useful because it can be applied to phenomena from different disciplines exhibiting the same type of phase transition. Kolmogorov-Johnson-Mehl-Avrami (JMAK) theory describes the decay of a metastable system to a unique equilibrium phase. This decay is driven by the difference between the free energy densities of the metastable and the equilibrium phase. The nucleation processes occur randomly in the metastable phase and the regions grow freely. The degree of phase transformation and in our case the fraction of volume transforming from one phase to another, f(t), is described by:

$$f(t)=1-\exp(-Kt^{n}) \qquad\qquad [1]$$

where t is the time, K is the temperature dependent crystallization rate constant and n – the Avrami exponent - represents growth and nucleation behavior. This equation has been applied to almost all areas where first order transitions occur, including liquid crystals and polymers, but also in cosmology, biophysics, magnetism, and surface science [5, 7]. It has been used in Chemistry to calculate chemical crystallization rates and in Ecology [5]. The equation can be linearized as:

$$\ln[-\ln(\,1-f(t)\,)]=\ln K+n\ln t \qquad\qquad [2]$$

where, $\ln K$ is the y-intercept, and n is the slope of the fitted straight line. The theory has been extended from time evolution of the order parameter for first order phase transitions to two point correlation functions, enabling it to predict and explain results from small angle scattering [8]. Ramos applies this model to kinetic Ising lattice gas model [5]. In this model the elementary kinetic processes act on microscopic length and time scales, making it possible to explain nucleation and growth rates. Gough derives a more general equation from the logistic curve where the JMAK theory is only a special case [9].

MATERIALS AND METHODS

iPP and CNTs
Isotactic polypropylene was obtained from Scientific Polymer Products, Inc. in powder form as Catalog #130. Mutli-walled carbon nanotubes came from MER Corporation. MWCNT's were produced by chemical vapor deposition (CVD). Catalytic CVD MWNTs had 140 ± 30 nm diameters and 7 ± 2 micron length, with purity of >90% (Catalog # MRCSD).

Purifying MW Carbon Nanotubes by Oxidation
One gram of the MWNT's was suspended in a mixture of concentrated sulfuric acid and nitric acid (3:1 vol. ratio). This solution was sonicated in a Misonix water bath sonicator for 24 hrs at 50°C. The resultant suspension was diluted with 400 mL of deionized water and filtered through a 400 nm pore membrane (PTFE) until the water passing through the filter had a pH between 6 and 7. The dispersions were subsequently filtered or diluted to the desired concentration. The resulting MWNT's had a pH of between 3 and 3.5 and were stable. Atomic Force Microscopy (AFM) analysis shows that the CNTs were shortened between 30% and 40% of their original length, as expected.

Preparation of iPP/CNT Nanocomposites
Nanocomposites were prepared by sonicating MWCNT (2.97mg +/- .01mg) in Xylene (75ml) at 50°C in a flask for 30min. IPP (2.9g) was added to the Xylene containing the MWCNT to form ~0.1% by mass composite. The solution was heated and stirred by placing the flask in an oil bath at 130°C sitting on a hot plate. Once the iPP is completely dissolved, we poured the solution slowly into a non-solvent (a polar compound, for example alcohol) with about a 5x volume dilution. The resulting powder precipitate was dried and then pressed into a film using a compression molding hot press at temperature of 200°C and pressure of 3-6000 lbs/sq in.

Differential scanning calorimetry
DSC analysis was performed on the TA Instruments calorimeter, Model # Q100.

Thermal treatment
The iPP nanocomposites were melted for two minutes at 200°C to erase any thermal history, and then cooled at 10°C/min to 135°C where they were held for various times in order to complete the isothermal crystallization for the nanocomposites with various CNTs content.

Data processing
We used the widely accepted Avrami model to analyze the DSC data for isothermal procedures. By plotting the relative crystallinity, x, as ln[-ln(1 – x)] vs. ln$t$, from eq [2], the slope of the curve, n, was determined over the range where a linear relationship exists between the plotted parameters. We obtain the value of K by $K=\exp(b)$, where b is the y-intercept of the linear fit of the plot [10,11,12]

RESULTS AND DISCUSSION

Figure 1 shows the heat flow vs. time during isothermal crystallization at 135°C, for iPP/CNT nanocomposites, at the indicated compositions. Time to maximum heat flow is a strong function of the CNT content, indicating that the crystal growth is accelerated by the addition of CNTs.

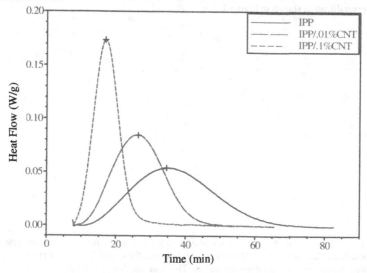

Figure 1. Heat flow vs. time for crystallization at 135°C for different iPP nanocomposites, as shown on the legend. The time to maximum heat flow, t_p, decreases systematically with CNT addition, and these times are listed in Table 1.

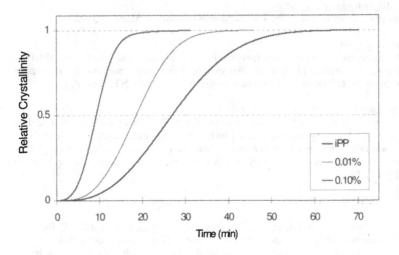

Figure 2. Relative crystallinity versus time from integrated heat flow data for the 135°C isothermal crystallization shown in Figure 1.

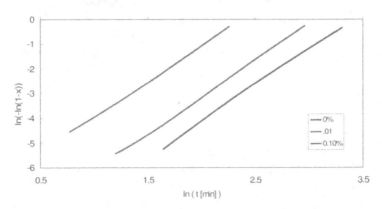

Figure 3. Avrami plots of $\ln[-\ln(1-X)]$ versus $\ln(t)$. The y intercept is proportional to the rate of crystallization K and the slope depends on the nucleation and the dimensionality of crystal growth, which determines the Avrami exponent n. The data are shown in Table 1.

As shown in Table 1, during the isothermal crystallization at 135°C the crystallinity is almost constant, but the crystallization time decreases dramatically, which is reflected in the increase of the value of K. The Avrami exponent n is a function of the dimensionality of growth: n between 1 and 2 indicates 1D growth; n between 2 and 3 represents a 2D growth; and n values between 3 and 4 indicates 3D growth [6] for heterogeneous nucleation processes. n also depends on the nucleation mechanism. The dimensionality reflected in the Avrami exponent n is also almost unchanged, and is around 3, which is typical for heterogeneous nucleation and three dimensional spherulitic growth. The crystal growth changes from spherulitic to fibrillar, originating on the surface of the nanotubes when CNTs are introduced [13, 14, 15]. The peak time dramatically deceases with adding nanotubes and the FWHM is also decreased greatly, pointing to a more uniform crystal population. The crystallization half time, for the crystallinity to reach 50% of the maximum crystallinity of the sample - $\tau_{1/2}$, is always longer than t_p, indicating the heat flow curves are asymmetric.

Table 1 – DSC Parameters[#] characterizing the rate of isothermal crystallization and the overall crystallinity of iPP/CNT nanocomposites.

iPP/CNT Nano-composite	H_C (J/g)	t_p (min)	FWHM (min)	X_c	n	K (1/min)	$\tau_{1/2}$ (min)
0% CNT	87.4	27.8	25.9	0.42	2.92	0.03	26.9
0.01% CNT	91.3	19.4	17.5	0.44	3.03	0.05	18.5
0.10% CNT	92.2	10.1	8.26	0.44	2.92	0.09	9.3

[#] The parameters are as follows: H_c is the integrated total exotherm area; t_p is the peak time, or the time to the maximum of the heat flow exotherm; FWHM is the Full Width at Half Maximum of the crystallization exotherm peak; X_c is the crystallinity mass fraction, determined by DSC as: $X_c=(H_c)/(H_f)$, where H_f is the heat of fusion for 100% crystalline iPP (H_f=209 J/g) [15]; n is the Avrami exponent, capturing the nucleation and dimensionality of crystallization; K is the crystallization rate constant with dimensions of 1/time, measures the rate of the crystallization process; n and K are determined from eq. 2; $\tau_{1/2}$ is the time for crystallization to reach 50% of the maximum crystallinity for the sample.

CONCLUSIONS

The crystallization rate constant K [min[-1]] as determined from the Avrami analysis is much larger for the samples with nanotubes, which reflects the faster crystallization rates. The $\tau_{1/2}$ and the time to maximum heat flow in the crystallization exotherm, t_p, decrease dramatically. FWHM narrows with increase of CNT concentration, which suggests a more uniform distribution of the formed crystals, similar to observations from other authors [6]. The Avrami exponent n does not change with increasing of the concentration of CNTs.

ACKNOWLEDGMENTS

For support of this research, the authors thank: Assumption College for a Faculty Development Grant; The Natural Science Department at Assumption College for the summer student stipends; The Petroleum Research Fund, 44149-AC7; The National Science Foundation, Polymers Program of the Division of Materials Research, DMR-0602473, and MRI Program under DMR-0520655 for thermal analysis instrumentation.

REFERENCES

1. F. Ciardelli, S. Coiai, E. Passaglia, A. Pucci, and G. Ruggeri, Polymer International **57**, 805 (2008).
2. C.A. Avila-Orta, F.J. Medellýn-Rodrýguez, M.V. Davila-Rodrýguez, Y.A. Aguirre-Figueroa, K. Yoon, B.S. Hsiao, Journal Of Applied Polymer Science **106**, 2640 (2007).
3. D. Xu, Z. Wang, Polymer, **49**, 330 (2008).
4. L. Valentini, J. Biagiotti, M.A. López-Manchado, S. Santucci, J.M.Kenny, Polymer Engineering and Science, February **44**, 303 (2004).
5. R. Ramos, P.A. Rikvold, M.A. Novotny, Physical Review B, **59**, 9053 (1999).
6. B.P. Grady, F. Pompeo, R.L. Shambaugh, D.E. Resasco, J. Phys. Chem. B, **106**, 5852-5858 (2002).
7. K. Avramova, Cryst. Res. Technol. **37(5)**, 491 (2002).
8. K. Sekimoto, Physics Letters 5 **105A**, 8 (1984).
9. T. Gough, R. Illner, VLSI Design **9**, No. 4, pp. 377-383 (1999).
10. J. Sandler, G. Boza, M. Nolte, K. Schulte, Y. Lam, M.S.P Shaffler, J. of Macromolecular Science **B42**, 479 (2003).
11. G. Xu, L. Du, H. Wang, R. Xia, X. Meng, Q. Zhu, Polymer International **57**, 1052–1066 (2008).
12. J. Young Kim, H.S. Park, S.H. Kim, Polymer **47**, 1379 (2006).
13. G. Georgiev, Y. Cabrera, M. Cronin, C. Rocheleau, B. Feinberg, P. Cebe, Bulletin of the American Physical Society, **53(1)** R1.00132 (2008).
14. E. Assouline, A. Lustiger, A. H. Barber, C. A. Cooper, E. Klein, E. Wachtel, H. D. Wagner, Journal of Polymer Science: Part B: Polymer Physics **41**, 520 (2003).
15. C. A. Avila-Orta Francisco, J. Medellýn-Rodrýguez, M. V. Davila-Rodrýguez,Y. A. Aguirre-Figueroa, K. Yoon, B. S. Hsiao, Journal of Applied Polymer Science, **106**, 2640 (2007).

Al, 133

barrier layer, 85

calorimetry, 185
crystal growth, 49, 173, 185

electron irradiation, 167
epitaxy, 73, 139, 167

grain
    boundaries, 133
    size, 133

ion-beam
    assisted deposition, 3, 15, 25,
        37, 49, 59, 65, 73, 85, 91,
        111, 123
    processing, 65, 117
ion-solid interactions, 65

laser
    ablation, 151
    annealing, 179

nitride, 59, 85
nucleation and growth, 37, 179

organic, 139

polymer, 185

Si, 73, 173, 179
simulation, 173
superconducting, 15, 111, 117,
    123, 151

texture, 3, 15, 25, 37, 59, 91, 111,
    117
thin film, 3, 25, 49, 91, 123, 139,
    151, 167

Printed in the United States
By Bookmasters